Prenatal Genetic Counseling

Prenatal Genetic Counseling

Practical Support for Prenatal Diagnostics, Decision-Making, and Dealing With Uncertainty

Edited by

SAM RIEDIJK
Medical Psychology, Department of Clinical Genetics, Erasmus Medical Centre, Rotterdam, The Netherlands

KARIN E.M. DIDERICH
Department of Clinical Genetics, Erasmus Medical Centre, Rotterdam, The Netherlands

ELSEVIER

ACADEMIC PRESS
An imprint of Elsevier

Academic Press is an imprint of Elsevier
125 London Wall, London EC2Y 5AS, United Kingdom
525 B Street, Suite 1650, San Diego, CA 92101, United States
50 Hampshire Street, 5th Floor, Cambridge, MA 02139, United States
The Boulevard, Langford Lane, Kidlington, Oxford OX5 1GB, United Kingdom

British Library Cataloguing-in-Publication Data
A catalogue record for this book is available from the British Library

Library of Congress Cataloging-in-Publication Data
A catalog record for this book is available from the Library of Congress

ISBN: 978-0-12-823329-0

For Information on all Academic Press publications
visit our website at https://www.elsevier.com/books-and-journals

Publisher: Kayla Wolfe
Acquisitions Editor: Peter B. Linsley
Editorial Project Manager: Billie Jean Fernandez
Production Project Manager: Selvaraj Raviraj
Cover Designer: Hugo de Pagter

Typeset by MPS Limited, Chennai, India

Dedication

In loving memory of Mae, Noud, Loes

Contents

List of contributors

Hilmar H. Bijma
Department of Obstetrics and Gynaecology, Division of Obstetrics and Prenatal Medicine, Erasmus MC, Rotterdam, The Netherlands

Hennie T. Brüggenwirth
Department of Clinical Genetics, Erasmus Medical Centre, Rotterdam, The Netherlands

Frederike Dekkers
Psychotherapy, Private Practice, Rotterdam, The Netherlands

Karin E.M. Diderich
Department of Clinical Genetics, Erasmus Medical Centre, Rotterdam, The Netherlands

Fadua El Bouazzaoui
Foundation Health Educators, Rotterdam, The Netherlands

Iris Emons
Experience Expert, The Netherlands

Jane Fisher
Antenatal Results and Choices (ARC), London, United Kingdom

Iris Jansen-Bakkeren
Department of Clinical Genetics, Erasmus Medical Centre, Rotterdam, The Netherlands

Marieke Joosten
Department of Clinical Genetics, Erasmus Medical Centre, Rotterdam, The Netherlands

Jasmijn E. Klapwijk
Department of Clinical Genetics, Erasmus Medical Center, Rotterdam, The Netherlands

Maarten F.C.M. Knapen
Department of Obstetrics and Gynaecology, Division of Obstetrics and Prenatal Medicine, Erasmus MC, Rotterdam, The Netherlands

Celine Lewis
Population, Policy and Practice, UCL Institute of Child Health, London, United Kingdom

Stina Lou
Center for Fetal Diagnostics, Aarhus University Hospital, Aarhus, Denmark; DEFACTUM – Public Health & Health Services Research, Central Denmark Region, Aarhus, Denmark

Kelly E. Ormond
Department of Genetics and Stanford Center for Biomedical Ethics, Stanford School of Medicine, Stanford, CA, United States

Eva Pajkrt
Department of Obstetrics and Gynaecology, Amsterdam UMC, University of Amsterdam, Amsterdam, The Netherlands

Ingrid A. Peters
Obstetrics and Gynaecology, Erasmus University Medical Center, Rotterdam, The Netherlands; Quality and Patient Care Erasmus University Medical Center, Rotterdam, The Netherlands

Olav B. Petersen
Department of Obstetrics, Juliane Marie Centre, Copenhagen University Hospital – Rigshospitalet, Copenhagen, Denmark; Department of Clinical Medicine, University of Copenhagen, Copenhagen, Denmark

Lara Ras
Experience Expert, The Netherlands

Sam Riedijk
Medical Psychology, Clinical Genetics Erasmus Medical Centre, Rotterdam, The Netherlands; Department of Clinical Genetics, Erasmus Medical Centre, Rotterdam, The Netherlands

Malgorzata I. Srebniak
Department of Clinical Genetics, Erasmus Medical Centre, Rotterdam, The Netherlands

Liesbeth van Leeuwen
Department of Obstetrics and Gynaecology, Amsterdam UMC, University of Amsterdam, Amsterdam, The Netherlands

Ida Vogel
Department of Clinical Genetics, Aarhus University Hospital, Aarhus, Denmark; Center for Fetal Diagnostics, Aarhus University Hospital, Aarhus, Denmark

Hajo I.J. Wildschut
Wildschut Consultancy, Enkhuizen, The Netherlands

Diana Zavyalova
Minor "Genetics in Society", Erasmus Medical Center, Rotterdam, The Netherlands

Preface

Karin E.M. Diderich and I have been working closely together in the prenatal genetics clinic of our hospital for the past 15 years. Over the years, we have worked with many couples who have faced fetal anomalies, and our longitudinal experience has provided us with valuable insights that we often use during our counseling sessions. Couples are often most supported by hearing from us how other couples reach their decisions and what they experience, providing them with a frame of reference. We increasingly realized that our experience is not easily acquired but is of great value. We, therefore, did not hesitate when Elsevier approached us to write a professional handbook for prenatal genetics. A team of coauthors was easily assembled, but we realized that our way of working would be much better reflected if we were to cocreate with our patients and even our students. Therefore we invited Lara and Iris, the mothers of Mae and Noud, respectively, to contribute. Both Lara and Iris shared their experiences with prenatal diagnosis in our education and were as motivated as we were, to share their experiences in this book. Hugo, Lara's partner and father of Mae, has designed the cover of this book, conveying the feeling they had after hearing the bad news.

The scientific literature, our experience, and this cocreation illustrate that care for couples in prenatal genetics comes straight from the heart. We hope that health-care professionals around the globe will find support and insights to support couples who face the most difficult times of their lives.

SECTION I

The Landscape of Prenatal Diagnosis

CHAPTER 1

A flow from screening to diagnostics

Ida Vogel[1,2], Stina Lou[2,3] and Olav Bjørn Petersen[4,5]
[1]Department of Clinical Genetics, Aarhus University Hospital, Aarhus, Denmark
[2]Center for Fetal Diagnostics, Aarhus University Hospital, Aarhus, Denmark
[3]DEFACTUM – Public Health & Health Services Research, Central Denmark Region, Aarhus, Denmark
[4]Department of Obstetrics, Juliane Marie Centre, Copenhagen University Hospital – Rigshospitalet, Copenhagen, Denmark
[5]Department of Clinical Medicine, University of Copenhagen, Copenhagen, Denmark

Denmark was one of the first countries in the world to introduce a universal and free-of-charge offer of combined first-trimester screening (CFTS) for chromosomal anomalies to all pregnant women. In the present chapter, we describe the development and current status of the program and reflect on future directions in the era of exome sequencing (ES) and cell-based noninvasive testing. Now, prenatal screening programs differ vastly all over the world and even between countries that are often considered comparable, such as Denmark and Norway. In this chapter, we will present a few such comparisons but will focus mainly on a thorough description of the Danish context. The aim is to give the reader a point of reference to consider and compare their specific, national context.

Introduction of combined screening

In 2000 the Danish Board of Health commissioned a medical working group (consisting of doctors and midwives) to prepare a possible revision of the then-existing guidelines for prenatal testing in Denmark which consisted of an offer of invasive testing for women of the age of 35 or above (Knudsen, 1994). One motivation was to counteract emerging inequality in health due to local discrepancies, where ultrasound examinations and prenatal screening were offered only in some counties (Lou et al., 2018). The work resulted in a significant change and new guidelines that were organized around the principle of informed choice. It introduced the offer of a CFTS for chromosomal anomalies and a second-trimester malformation scan to all pregnant women irrespective of age and residence. The main advantages of shifting from age-related screening to universal, first-trimester screening, were argued to be a higher predictive value compared to the previous regime and an expected decrease in the number of invasive tests and procedure-related miscarriages [estimated at the time to be about 1% (Tabor et al., 1986)]. The official aim stated in the guidelines was to facilitate individual choice:

> *The aim of prenatal testing is – within the juridical framework of Danish Law – to assist a pregnant woman, if she wants such assistance, in making autonomous, reproductive*

Prenatal Genetic Counseling
DOI: https://doi.org/10.1016/B978-0-12-823329-0.00007-6

decisions. Neutral and adequate information is a necessary condition to this end.... The aim of prenatal testing is not to prevent the birth of children with serious diseases or handicaps (Parliamentary Decision on Prenatal Diagnosis, May 15, 2003).

After a formal process of public hearing where patient organizations and central agencies were invited to express their opinions, in September 2004 the new guidelines on prenatal testing in Denmark were published (Sundhedsstyrelsen, 2004). This meant in practice that CFTS was introduced as an option for all pregnant women in Denmark.

Implementation of CFTS, effects, and public discourse

All five Danish counties were quick to follow the new guidelines and by June 2006, the whole of the country was covered (Ekelund et al., 2008). However, implementing the guideline did represent a considerate, collective challenge for all departments with regards to increased requirements of equipment, staff, and education. In hindsight, these collective challenges served as a common platform, for example, for developing educational standards and for the procurement of equipment, which further developed collaboration within clinical genetics, fetal medicine, and across obstetric ultrasound units. Furthermore, the clinical collaborations following the 2004 guideline gave the medical specialties a joint understanding of the importance of a high-quality, uniform clinical practice to ensure a consistent and equal offer of prenatal screening, testing, and care to all pregnant women in Denmark. In comparison, Sweden still has considerable differences in the prenatal screening offered in the different counties (Petersson et al., 2016; Ingvoldstad et al., 2014).

Right from the start, the screening uptake was—and has remained—very high (>90%) (Kirkegaard et al., 2008; RKKP, 2011). Today, many Danish women consider the CFTS an integral part of standard prenatal care (Lou et al., 2017; Schwennesen and Koch, 2012). Following the introduction of CFTS, the number of invasive procedures decreased markedly while the positive predictive value rose (Fig. 1.1). Concurrently, there was an increase in prenatal diagnoses of Down syndrome (DS) (Lou et al., 2018) (Fig. 1.2). Additionally, the number of DS live births decreased suddenly and significantly but subsequently stabilized at 20–30 annual live births. Of these the majority were diagnosed postnatally, including women with a screen positive CFTS result who decided to continue pregnancy without additional testing.

Studies have shown that the majority of Danish women have a positive attitude toward CFTS and participate in screening based on an informed decision (Bangsgaard and Tabor, 2013; Dahl et al., 2011). However, the high uptake has led some to concern over routine participation and lack of informed choice. Particularly the significant decline in the number of children born with DS continues to be a debate in Danish (Lou et al., 2020) and international media (Zhang, 2020). An analysis of the Danish

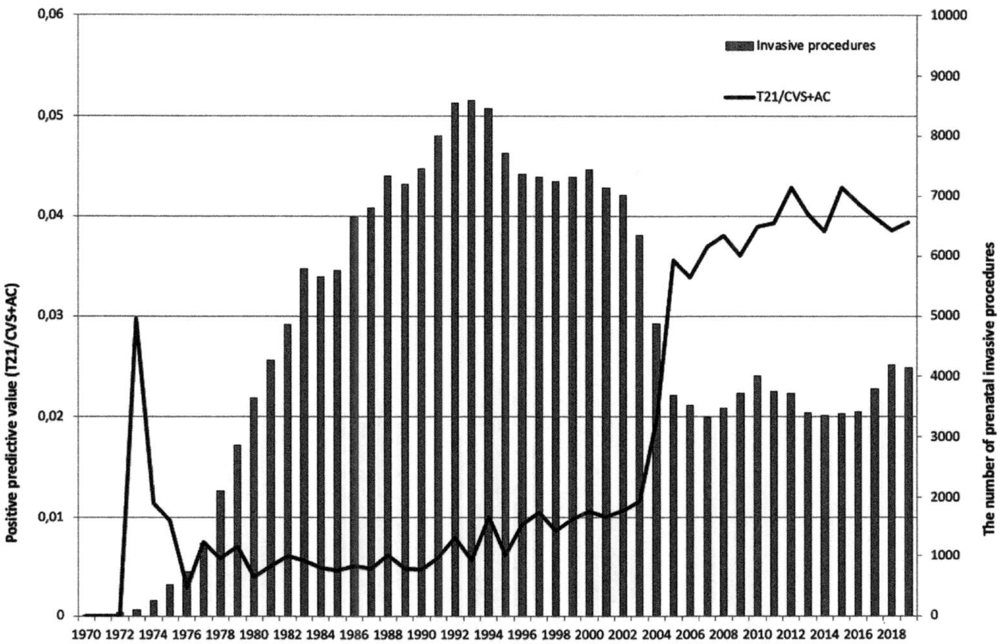

Figure 1.1 Number of invasive procedures and the aggregate positive predictive value defined as T21/invasive samples (all indications) in Denmark (not Greenland or the Faroes) 1973–2019. *Data from the Danish Central Cytogenetic Registry.*

media representations of the screening program and DS identified four discourses of which only one was somewhat positive toward the screening offer. Thus despite the high screening uptake and the high percentage of women who choose to terminate the pregnancy following a prenatal DS diagnosis, the legitimacy of the screening offer continues to stir debate in Denmark.

The discussion of chromosomal microarray, karyotyping, and noninvasive prenatal testing in Denmark

Already around 2005 the much-increased diagnostic rate of chromosomal microarray (CMA) over karyotyping on children with developmental delay or malformation was universally acknowledged and, thus, provided easy diagnostic opportunity for many more syndromes than DS. CMA can detect deletions and duplications down to 0.1 Mb compared to the 10-Mb detection limit of karyotyping, and CMA can hereby provide a one-shot diagnosis of all microdeletion and duplication syndromes (as, e.g., 22q11.2 deletion syndrome). Further, the waiting time for a result is 3–5 days for CMA compared to 10–14 days for karyotyping (Wapner et al., 2012; Vestergaard et al., 2013). Still, there was quite some hesitance in implementing CMA in prenatal

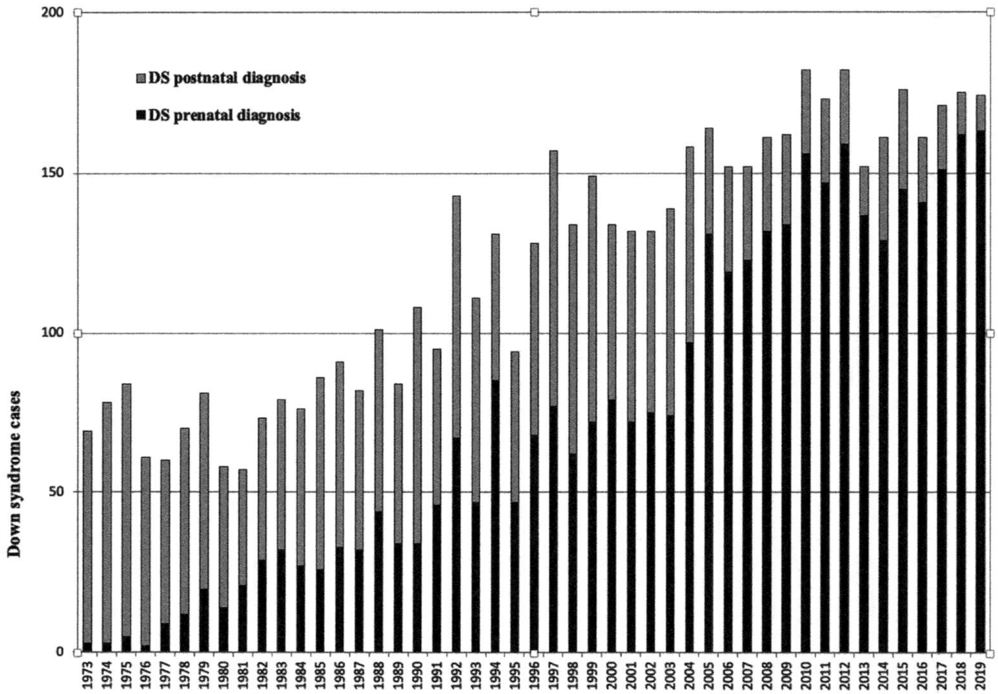

Figure 1.2 Number of Downs syndrome cases according to pre- or postnatal diagnosis in Denmark (not Greenland or the Faroes) 1973–2019. *Data from the Danish Central Cytogenetic Registry.*

diagnosis due to a fear of findings of unknown significance. However, for the analyses of invasive samples, an emphasis was put not only on the ability to detect more serious disorders but also on the shorter waiting time for a result. Both these points accelerated the change from karyotyping to CMA in spite of concurrent counseling difficulties with rare variants of unknown significance and susceptibility loci. However, the change was not made simultaneously between Danish laboratories, and even though CMA was the standard analysis in one region from 2015, only in 2020 were all regions offering CMA as the first-tier analysis. Several Danish regions have now published own results on the increased resolution of CMA over karyotyping on various indication, and here in 2020 this discussion weaned out, and all adhere to a national guideline suggesting CMA for all invasive samples (DFMS, 2018).

Internationally, noninvasive prenatal testing (NIPT) was getting increasingly popular from 2012 and onward, and several publications came out documenting decreases in the invasive rate after an NIPT offer was available (Larion et al., 2014; Chan et al., 2015; Chetty et al., 2013). Other countries have implemented NIPT as the first-line analyses for all pregnant women (van der Meij et al., 2019; Gadsboll et al., 2020). In Denmark the discussions on the use of NIPT versus invasive samples with CMA for different indications were also ongoing in both genetic and fetal medicine networks (Petersen et al., 2014; Vogel et al., 2018). It played a central role in these discussions that recent Danish data documented that the risk of doing invasive testing was lower than expected previously (Wulff et al., 2016). As a result, the Danish medical societies on medical genetics and fetal medicine found it easier to recommend invasive procedures. Furthermore, one of the very early Danish NIPTs turned out to be one of the rare false negatives (iso21) (Huijsdens-van Amsterdam, et al., 2018), and this case had a significant negative impact on the impression of NIPTs' usefulness for screening purposes among health practitioners. Behind these discussions, there was a substantial concern among health providers for equal access to healthcare across regional borders and social groups, and NIPT quickly became accessible in private clinics, but the volume turned out to be low (Lund et al., 2021). NIPT has been used publicly since 2017 as an alternative to invasive testing but only for trisomy 13, 18, and 21 and sex, and, thus, resolution is lower than karyotyping. Additionally, NIPT has a longer turnaround time in the regional setups in Denmark than CMA on invasive sampling due to low volumes and a need to bulk samples from several days (Lund et al., 2021). As a result of this, there has been no decrease in the invasive rate in Denmark and less than 1% of all pregnancies are tested with NIPT.

Current status

In 2017 the Danish Board of Health revised the guidelines to include an offer of NIPT as an alternative to invasive testing or no testing for women with a high-risk CFTS result (Fig. 1.1). Contrary to other countries, the majority of Danish women with a high-risk CFTS result still choose the invasive test (Lund et al., 2021). One explanation is that the medical societies have been very vocal in their recommendation of the invasive procedure due to better diagnostic performance (Petersen et al., 2014) and the recent evidence that the risk of miscarriage following an invasive procedure is significantly lower than previously estimated (Wulff et al., 2016).

Another recent development is that we have seen an annual increase in the number of children born after a prenatal diagnosis of DS (Fig. 1.3) (Lou et al., 2018). The National Down Syndrome Association was quick to interpret this as a sign of more women actively choosing to continue a pregnancy after a DS diagnosis and the media ran this story. However, the total number of children born with DS has not increased. Rather, the increase should be interpreted as a result of the option of NIPT to women, who would previously have declined invasive procedures (Miltoft et al., 2018). By opting for risk-free testing, these couples may potentially receive specialized care for their DS fetus/baby (by a cardiologist, additional surveillance during labor, checkup by a specialized pediatrician immediately after birth); however it is yet to be studied how this subgroup applies these options and whether they have any effects.

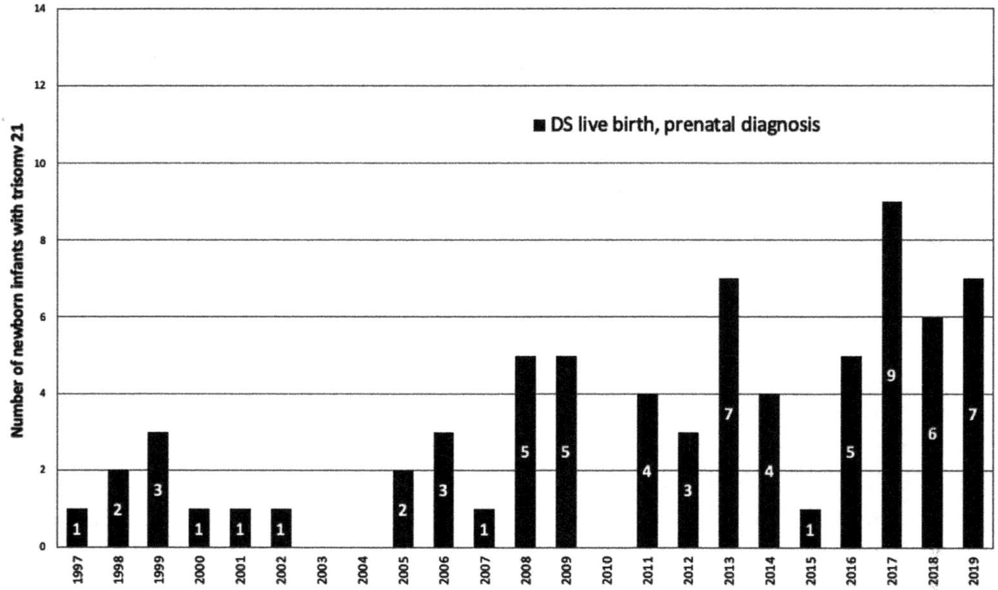

Figure 1.3 Number of newborn infants diagnosed with T21 before third trimester in Denmark 1997–2019. *Data from the Danish Central Cytogenetic Registry.*

The flow from screening to diagnosis

The general practitioner as point of entry

When a woman has a positive pregnancy test, she will book an appointment at the general practitioner (GP) for the verification of pregnancy, for blood sampling, and a referral to relevant services of maternity care. It is the GP who is responsible for informing the woman about the offer of prenatal screening and making sure that she makes an informed choice about potential participation. This setup has been continuously critiqued, as both qualitative studies and clinical experiences indicate that the GPs have often not discussed the screening and potential implications in much detail. Thus women do not always consider prenatal screening an active choice. This is in concordance with findings from several other Western countries (Gottfredsdottir et al., 2009; Han, 2013).

The first-trimester ultrasound scan and combined screening

If the woman chooses CFTS, then blood samples (PAPP-A and ßhCG) are drawn and analyzed prior to the ultrasound examination. The first-trimester ultrasound is performed at the obstetric department in regional/university public hospitals, by nurses or midwives all certified in sonography by the Fetal Medicine Foundation, which

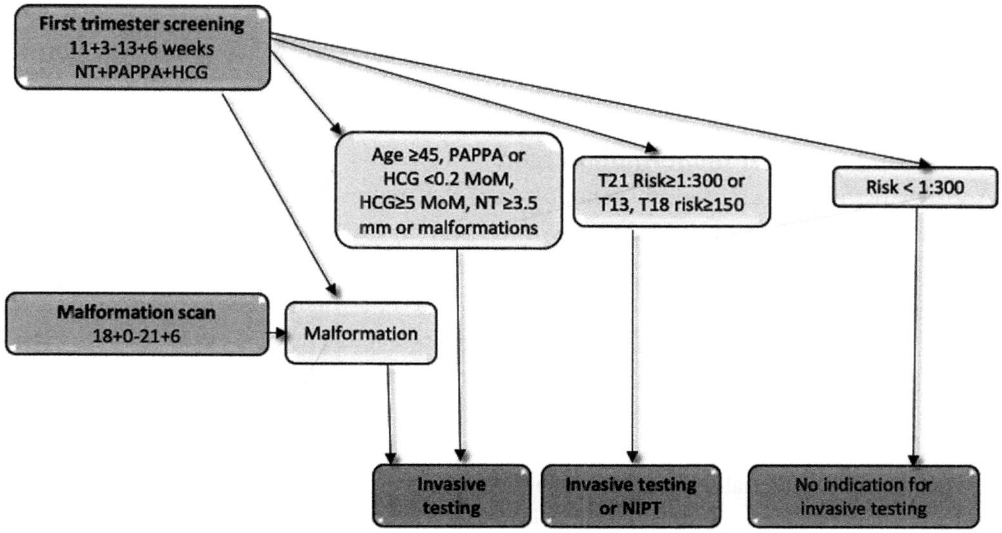

Figure 1.4 The national prenatal screening program in Denmark 2017 and onwards.

includes annual audit and recertification. Prior to starting the examination, consent for CFTS should be verbally confirmed to the sonographer.

Following the ultrasound scan, the CFTS is calculated on the spot based on age, nuchal translucency, PAPP-A, and free ßhCG, and the results are delivered to the couples by the sonographer. In some departments the actual risk number is conveyed for all, while in other departments the risk is only conveyed for the high-risk group. In the case of a high-risk CFTS (Fig. 1.4), some hospitals will have a fetal medicine specialist to explain the result, but in most cases it is the sonographer who provides information about the result and the options available, and—if necessary—supports the couple during their process of decision-making. The rationale behind letting sonographers do the counseling is that it is less anxiety-provoking for the couple. It is a highly stressful situation while the vast majority of cases with an increased risk are normal.

About 5% of the high-risk pregnancies choose no further diagnostics; however, of those who choose further diagnostics, 80% opt for invasive testing with CMA analysis and only 20% opt for NIPT (Lund et al., 2021).

Turnaround time and delivery of results

It is well documented that a high-risk screening result leads to a significant increase in worry and anxiety in pregnant women (Lou et al., 2016). As time is essential for the couples' coping, we make short turnaround time a priority. For the vast majority (>90%) (Vogel et al., 2018), the analysis shows a normal result and from the international literature (Lou et al., 2015), we know that the anxiety levels among these women subsequently drop to the same level as screen negative women.

Diagnoses are delivered by phone (by the obstetrics department) with an offer of follow-up at the Department of Clinical Genetics. In Denmark, all prenatal counseling is performed by clinical geneticist consultants or residents.

The second-trimester malformation scan

Another pathway to a prenatal genetic test is the malformation scan in second trimester, and 96% of all Danish pregnant women participate. If malformations are detected, then invasive sampling is offered and almost always accepted (Fig. 1.4). Here only CMA is perceived a relevant analysis and not NIPT, as malformations detected at this point often are due to syndromes that are not currently picked up by NIPT. However, when an anomaly or malformation is detected in the second or third trimester, CMA offers an explanation in less than one in five cases. This means that prospective parents may need to decide to terminate or continue a pregnancy without knowing if the malformation is a symptom of a multiorgan syndrome, a treatable disease, or a single malformation. Furthermore, in ongoing pregnancies, the prenatal surveillance, the delivery, and the postnatal treatment and care are carried out without knowing the specific diagnosis/prognosis of the child, which often leads to over- or undertreatment of the child, failed or ineffective therapeutic interventions, parental anxiety, as well as uncertainty and frustration among health professionals. It is hoped that the introduction of ES may provide a higher diagnostic yield for this group. In Denmark, indications and setups for ES of these pregnancies are currently being trialed locally and soon the guideline process of evening out differences will commence.

Termination of pregnancy

The stated aim of the prenatal screening program is to provide prospective parents with reproductive choices, which includes the decision to the termination of pregnancy (TOP). In Denmark, all women have the legal right to TOP up to week 11 + 6. From week 12 + 0, TOP is only allowed after application to a regional council, permission will be granted if (1) there is a risk of severe fetal disease, or the newborn will suffer from severe physical or mental disorder, or (2) if the pregnancy or care of the child will pose a risk to the mother's physical or mental health, or (3) because of social circumstances. Annually, there are 60,000 live births, and 500 abortions after week 12 due to severe physical or mental fetal disorder (RKKP, 2019). Often, the medical clinical geneticist and other staff will need to guide and support the couples during this difficult decision-making process. TOP in the third trimester is very rarely granted, and, thus, there has been a reluctance toward prenatal diagnoses in late pregnancy and a preference for postponing the diagnosis until after birth. However, recently—and propelled by the introduction of ES—we have become more aware of the value of a prenatal diagnosis even in the absence of TOP, as it allows for the couple to adjust to this information, to prepare for birth and specialized postnatal care.

Data in registries

The prenatal screening setup in Denmark is characterized by a high level of cross-regional collaboration, continuous interdisciplinary dialog between the medical societies for fetal medicine and for clinical genetics, as well as collective commitment to gathering and sharing data. As a result, Denmark has high-quality databases combining cytogenetic analyses (Danish Central Cytogenetic Registry) with prenatal ultrasound (Danish Fetal Medicine Database) with data on all pregnancies from week 12, including spontaneous abortions, TOPs, and live-born babies (Ekelund et al., 2016; Nguyen-Nielsen et al., 2013). This collaboration and focus on high-quality data is a result of the standardization of prenatal screening initiated with the publication of 2004 national guidelines. The Danish Fetal Medicine Database enables close and consistent monitoring of a wide range of quality parameters to assess performance (e.g., false-positive rates), allow benchmarking between hospitals, and support continuous quality improvement. This availability of current and precise data has proved very valuable for generating context-specific knowledge (e.g., on the lower risk of procedure-related miscarriage) and for assessing new technologies' potential consequences in a Danish context (e.g., the value of CMA over NIPT). Our international readers may be interested to know that data can be made available to all by application.

International comparisons

The UK context in Table 1.1 *is written by Jane Fischer, Antenatal Results and Choices that is a UK-based charity providing impartial information and support to women and couples through antenatal screening and its consequences.*

There are tremendous differences between many countries regarding scope, content, and uptake of the public prenatal screening programs (Gadsboll et al., 2020). For comparison the screening context in selected European countries is presented in Table 1.1. In the Netherlands the scope for the national screening program was to offer testing for T13, 18, and 21 (van der Meij et al., 2019), whereas it in Denmark is to detect "serious disorder or handicap," and the differences in these scopes may adequately explain why the countries have implemented NIPT and CFTS, respectively.

Table 1.1 Population and reproductive data as well as essential data on legislation and prenatal screening for comparison of selected European countries

Country	Population (in Mio; 2018)	Annual births (in 1000; 2018)	Annual TOP (in 1000; 2017/18)	TOP legislation	National guideline/program for first-trim PND	Scope /chromosomal anomalies covered by national program	Uptake of national program (%)	References
The United Kingdom	67.1	730.9	213.8	Abortion law in England, Scotland, and Wales allows pregnancies to be legally terminated providing that two doctors certify that an appropriate ground has been met. Ground E of the Act was designed to cover prenatal diagnosis and enables termination without gestational time limit providing two doctors certify that *there is substantial risk that if the child were born it would suffer from physical or mental abnormalities as to be seriously handicapped*. In the context of survivable fetal anomalies, some doctors sanction the termination under Ground C which stipulates that *the continuance of the pregnancy would involve risk to the life of the pregnant woman, greater than if the pregnancy were terminated*. The gestational limit for Ground C is 24 weeks, and over 97% of all abortions in Britain are performed under this clause. Over the last decade, there has been an increase in prenatal diagnosis emanating from first-trimester screening, but still many women receive news from the mid-pregnancy scan that leads to decisions about the future of the pregnancy having to be made close to the 24-week limit for most abortions. And as more	In the United Kingdom the majority of pregnant women access maternity care through the publicly funded NHS. In England, Scotland, and Wales, there is a standardized screening program for sickle cell and thalassemia, Down's syndrome, Edwards' syndrome, and Patau's syndrome and a specified "menu" of structural conditions. The UK National Screening Committee makes recommendations on national screening programs NHS screening (2018). Screening is optional, and pretest information and counseling aims to help women and partners to make a "personal informed choice" whether to accept or decline the offer	Trisomy 13, 18, and 21	77.1	NHS screening (2018), UK. Legislation (1967), Legislation.Gov. UK (1990), NHS Maternity Statistics (2019), National Statistics (2019)

				women are having third-trimester scans to monitor fetal growth or placental position, a significant minority faces a diagnosis and termination of pregnancy in the third trimester				
Poland	37.9	388.2	1.8	Up to November 2020: Allowed up to point of postnatal viability when the pregnancy poses a threat to the life or health of a woman, or there is a high probability of severe and irreversible impairment of the fetus or an incurable disease threatening its life. And allowed up to 12 weeks when the pregnancy is the result of a prohibited act (rape or incest). A new legislation is in process, planned start November 2020, restricting TOP to women where the pregnancy is the result of a prohibited act (rape or incest, up to 12-week gestation), or if the pregnancy poses a life-threatening risk of the woman	The Polish Society for Ob Gyn recommend that all women get a first-trimester ultrasound scan according to Fetal Medicine Foundation standards, including combined screening for chromosomal anomalies based on maternal, biochemical, and ultrasound markers. And offer invasive testing (CVS) for women at high risk (>1:300) for trisomy 21. And NIPT is recommended for the intermediate risk group (1:300–1:1000) The public healthcare reimburse costs for these groups (one or more criteria): over 35 years of age, a history of chromosomal aberration (previous pregnancy or prospective parents), other background for high risk of genetic- or syndromic child, or abnormal ultrasound or biochemical tests in current pregnancy	None specified, in practice all detectable chromosomal anomalies	Data not available	Sas (2020), Wikipedia (2020), Kosinski, et al. (2019), Szczurek-Żelazko (2018), Borowski, et al. (2020)
The Netherlands	17.1	168.5	30.1	Allowed up to 24 weeks. There is an obligatory 5-day period for reflection between the first visit to the doctor and the procedure itself	All women are offered first-trimester screening by the TRIDENT-2 protocol, which includes (1) early ultrasound for the determination of GA/EDD, and (2) a choice between either 2a NIPT or 2b first-trimester combined	Trisomy 13, 18, and 21	43	Termination of Pregnancy in the Netherlands (1984), Crombag et al. (2014), Garssen and Kruijer (2011), Oepkes et al. (2016), van der Meij et al. (2019)

(*Continued*)

Table 1.1 (Continued)

Country	Population (in Mio; 2018)	Annual births (in 1000; 2018)	Annual TOP (in 1000; 2017/18)	TOP legislation	National guideline/program for first-trim PND	Scope /chromosomal anomalies covered by national program	Uptake of national program (%)	References
					screening for chromosomal anomalies based on maternal, biochemical, and ultrasound markers. Costs are partly reimbursed, self-pay amounts 177€			
Belgium	11.4	118.3	17.1	Allowed up to 14 weeks. There is an obligatory 6-day period for reflection between the first visit to the doctor and the procedure itself. After 14 weeks, only by application in the case of pregnancy it endangers the women's health, or if the baby will likely suffer from an incurable and serious condition. There is currently an ongoing political process aiming to change the 14-week limit to 18 weeks, and the 6-day limit to 2 days	All women are offered first-trimester screening, including ultrasound for assessment of EDD and gross anomalies, and (since 2017) an NIPT. Costs are reimbursed, except for a minor self-pay of 9€	Trisomy 13, 18, and 21, and more, depending on NIPT, several providers offer WGS NIPT	78.9	Angoloinfo (2020), Neyt et al. (2014), Statista (2020)
Sweden	10.0	115.8	35.8	Allowed up to 12 weeks, and in practice always allowed up to 18 weeks after application. Thereafter only allowed by application up to 22 weeks in the case of (1) risk of severe fetal or neonatal/child physical or mental disease, or (2) if the pregnancy or childcare poses a significant risk of physical	All women are by national guideline recommended free-of-charge first-trimester screening, including ultrasound and biochemical test (KUB test), and NIPT or CVS + PCR in the case of high-risk. Implementation of this program varies between the 21 Swedish counties	Trisomy 13, 18, and 21	Uptake varies between regions and counties, from <10% to >90%	Heino and Gissler (2019, 2020), Malcus and Conner (2017), Petersson et al. (2016), Sverige_ Socialstyrelse (2018), Sveriges_ Riksdag (1990)

				or mental disease to the mother				
Denmark	5.7	61.5	14.3	Allowed up to 12 weeks, thereafter, allowed by application up to limit of viability, in practice up to 23 weeks, in the case of (1) if there is a risk of severe fetal disease, or the newborn will suffer from severe physical or mental disorder, or (2) if the pregnancy or care of the child will pose a risk to the mother's physical or mental health, or (3) because of social circumstances. After 23 weeks, only in the case of lethal fetal/child conditions	According to national guideline, all pregnant women are offered public free for all first-trimester ultrasound and combined screening, based on maternal characteristics, biochemical- and ultrasound markers. High risk defined as T21 >1:300, or T13/T18 >1:150, or single factors despite trisomy risk (maternal age, NT >3.5 mm, abnormal BC, or ultrasound findings associated with chromosomal or genetic disease)	Assessment of factors of importance for the health of the baby—including chromosomal anomalies. Risk is calculated for trisomy 13, 18, and 21, but assessment includes clinical relevant chromosomal anomalies	95	Heino and Gissler (2019, 2020), RKKP (2019, 2020), Abortankenævnet (2017)
Finland	5.5	47.6	8.7	Allowed up to 12 weeks (one doctor's approval needed). Thereafter only by application up to 20 weeks (two doctors' approval) or 24 weeks in the case of risk of fetal/child disease or pregnancy poses a maternal risk. After 24 weeks, only in the case of lethal fetal/child conditions	All women are offered public free for all first-trimester ultrasound and combined screening, based on maternal characteristics, biochemical- and ultrasound markers. CVS with CMA in the case of high risk (T21 >1:250)			Heino and Gissler (2019, 2020), THL_Finland (2020)
Norway	5.1	55.1	12.4	Allowed up to 12 weeks. Thereafter only by application (assessed by two doctors) up to 22 weeks in the case of risk of fetal/child disease or pregnancy poses a maternal risk. After 22 weeks, only in the case of lethal fetal/child conditions	Only women 38 year or older at EDD are offered free public first-trimester screening, including ultrasound and biochemical test (KUB test), and NIPT or CVS + PCR in the case of high risk (>1:250). There is an ongoing political process aiming first-trimester KUB screening to all women, and implementation of NIPT to high risk	Trisomy 13, 18, and 21	<10% of all pregnant women	Heino and Gissler (2019, 2020), Helse-Norge (2020), Løkeland et al. (2019), Helsedirektoratet Fosterdiagnostikk (2020)

(Continued)

Table 1.1 (Continued)

Country	Population (in Mio; 2018)	Annual births (in 1000; 2018)	Annual TOP (in 1000; 2017/18)	TOP legislation	National guideline/program for first-trim PND	Scope /chromosomal anomalies covered by national program	Uptake of national program (%)	References
Iceland	0.3	4.2	1.0	Allowed up to 12 weeks. Between 12 and 16 weeks allowed if two doctors agree on indication. Thereafter only by application up to 22 weeks in the case of risk of severe/lethal fetal/child disease or pregnancy poses a maternal risk. After 22 weeks only in the case of pregnancy poses a life-threatening maternal risk				Heino and Gissler (2019, 2020)

CMA, Chromosomal microarray; *NHS*, National Health Service; *NIPT*, noninvasive prenatal testing; *TOP*, termination of pregnancy.

When comparing countries with major differences in legislation regarding TOP, that is, the Northern European countries with Poland, where access to TOP is much more restricted, it is interesting that the uptake of prenatal screening in Poland exceeds 50%, and the uptake of NIPT is rapidly increasing. Thus different abortion legislations do not seem to explain national differences. Countries with comparable public funded screening programs, as the United Kingdom, Denmark, and Sweden have a huge variation in the uptake of screening—from <10% in some Swedish counties to 95% in Denmark. Even geographic and cultural neighbors as The Netherlands and Belgium that both use NIPT as first-line test have a significant difference in uptake in the use of NIPT 42% and 79%, respectively (De Zordo et al., 2021). Do the women living not far apart really request substantial different prenatal screening programs? Or do the politicians and health-care professionals not provide the service the women want? Or do factors like who performs the counseling, the national public, and social discourse or other factors matter more than anticipated? At present, there is poor understanding for these differences, and more research is needed to understand this.

The future

It is always hard to predict the future. However, we anticipate four things (1) that the differences between screening and diagnostics will become more overlapping as NIPT develops. (2) That the increased use of whole genome−based analyses (e.g., ES) will generate more unexpected/incidental and more uncertain findings and, thus, new forms of consent will need to be developed to balance patient request against unreadable legal documents. (3) Testing will become more prevalent in the third trimester as we learn to diagnose with the aims of birth and postnatal treatment. (4) We will need ways to allow data transfer from maternal patient files to the newborn ones so the prenatal history, including genetic analyses with potential lifelong relevance, is not lost for the newborn.

References

Abortankenævnet: *Årsberetning 2017, Abortankenævnet*, 2017, Styrelsen for Patientsikkerhed.

Angoloinfo: *Termination of pregnancy − abortion in Belgium*, 2020. https://web.archive.org/web/20160701223511/ https://www.angloinfo.com/brussels/how-to/page/brussels-healthcare-pregnancy-birth-termination-abortion.

Bangsgaard L, Tabor A: Do pregnant women and their partners make an informed choice about first trimester risk assessment for Down syndrome, and are they satisfied with the choice? *Prenat Diagn* 33(2):146−152, 2013.

Borowski D, Pietryga M, Basta P, et al: Practice guidelines of the Polish Society of Gynecologists and Obstetricians − Ultrasound Section for ultrasound screening in uncomplicated pregnancy − 2020, *Ginekol Pol* 91(8):490−501, 2020.

Chan YM, Leung WC, Chan WP, Leung TY, Cheng YK, Sahota DS: Women's uptake of non-invasive DNA testing following a high-risk screening test for trisomy 21 within a publicly funded healthcare system: findings from a retrospective review, *Prenat Diagn* 35(4):342–347, 2015.

Chetty S, Garabedian MJ, Norton ME: Uptake of noninvasive prenatal testing (NIPT) in women following positive aneuploidy screening, *Prenat Diagn* 33(6):542–546, 2013.

Crombag NM, Vellinga YE, Kluijfhout SA, et al: Explaining variation in Down's syndrome screening uptake: comparing the Netherlands with England and Denmark using documentary analysis and expert stakeholder interviews, *BMC Health Serv Res* 14:437, 2014.

Dahl K, Hvidman L, Jorgensen FS, Henriques C, Olesen F, Kjaergaard H, et al: First-trimester Down syndrome screening: pregnant women's knowledge, *Ultrasound Obstet Gynecol* 38(2):145–151, 2011.

De Jonge H, Van Ark T, Blokhuis P: Termination of Pregnancy in The Netherlands. Government of The Netherlands 2021 [cited 2021 08/01/2021]; Available from: https://www.government.nl/topics/abortion.

De Zordo S, Zanini G, Mishtal J, Garnsey C, Ziegler AK, Gerdts C: Gestational age limits for abortion and cross-border reproductive care in Europe: a mixed-methods study, *BJOG* 128(5):838–884. doi:10.1111/1471-0528, 2021.

DFMS, 2018. Guideline for using prenatal chromosomal microarray by the Danish Fetal Medicine Society.

Ekelund CK, Jorgensen FS, Petersen OB, Sundberg K, Tabor ADanish Fetal Medicine Research Group: Impact of a new national screening policy for Down's syndrome in Denmark: population based cohort study, *BMJ* 337:a2547, 2008.

Ekelund CK, Kopp TI, Tabor A, Petersen OB: The Danish Fetal Medicine database, *Clin Epidemiol* 8:479–483, 2016.

Gadsboll K, Petersen OB, Gatinois V, Strange H, Jacobsson B, Wapner R, et al: Current use of noninvasive prenatal testing in Europe, Australia and the USA: a graphical presentation, *Acta Obstet Gynecol Scand* 99(6):722–730, 2020.

Garssen J, Kruijer H: *Annual number of abortions stable over the past decade*, CBS Statistics Netherlands, 2011.

Gottfredsdottir H, Bjornsdottir K, Sandall J: How do prospective parents who decline prenatal screening account for their decision? A qualitative study, *Soc Sci Med* 69(2):274–277, 2009.

Han S: Pregnancy in practice. In Tremayne S, Inhorn MC, Kreager P, editors: *Expectation and experience in the contemporary United States*. New York: Berghahn; p 206, 2013.

Heino A, Gissler M: *Induced abortions in the Nordic countries 2017*, 2019. https://thl.fi/en/web/thlfi-en/statistics/statistics-by-topic/sexual-and-reproductive-health/abortions/induced-abortions-in-the-nordic-countries.

Heino A, Gissler M: *Nordic perinatal statistics 2018*, 2020, THL Finland.

Helse-Norge: *Fetal diagnostics in Norway*, 2020, Helse Norge. https://www.helsenorge.no/en/underso-kelse-og-behandling/fetal-diagnostics/. (Accessed November 04, 2020).

Helsedirektoratet Fosterdiagnostikk, N. https://www.helsedirektoratet.no/tema/bioteknologi/fosterdiagnostikk

Huijsdens-van Amsterdam K, Page-Christiaens L, Flowers N, Bonifacio MD, Ellis KMB, Vogel I, et al: Isochromosome 21q is overrepresented among false-negative cell-free DNA prenatal screening results involving Down syndrome, *Eur J Hum Genet* 26(10):1490–1496, 2018.

Ingvoldstad C, Georgsson Ohman S, Lindgren P: Implementation of combined ultrasound and biochemistry for risk evaluation of chromosomal abnormalities during the first trimester in Sweden, *Acta Obstet Gynecol Scand* 93(9):868–873, 2014.

Kirkegaard I, Petersen OB, Uldbjerg N, Torring N: Improved performance of first-trimester combined screening for trisomy 21 with the double test taken before a gestational age of 10 weeks, *Prenat Diagn* 28(9):839–844, 2008.

Knudsen JL: *Prænatal genetisk information, rådgivning og undersøgelse: vejledning og redegørelse*, København, 1994, Sundhedsstyrelsen, 112 sider p.

Kosinski P, Ferreira JCP, Lipa M, et al: Preferences and expectations among Polish women regarding prenatal screening, *Ginekol Pol* 90(9):544–548, 2019.

Larion S, Warsof SL, Romary L, Mlynarczyk M, Peleg D, Abuhamad AZ: Uptake of noninvasive prenatal testing at a large academic referral center, *Am J Obstet Gynecol* 211(6):651–657, e1, 2014.

Legislation.Gov.UK, editors: *Human Fertilisation and Embryology Act 1990*, London, 1990, Legislation.Gov. UK.

Løkeland M, Akerkar R, Askeland OM, et al: *Rapport om svangerskapsavbrot for 2018 i Norge*, 2019, Folkehelseinstituttet.

Lou S, Lanther MR, Hagenstjerne N, Bjorn Petersen O, Vogel I: "This is the child we were given": a qualitative study of Danish parents' experiences of a prenatal Down syndrome diagnosis and their decision to continue the pregnancy, *Sex Reprod Healthc* 23:100480, 2020.

Lou S, Mikkelsen L, Hvidman L, Petersen OB, Nielsen CP: Does screening for Down's syndrome cause anxiety in pregnant women? A systematic review, *Acta Obstet Gynecol Scand* 94(1):15–27, 2015.

Lou S, Nielsen CP, Hvidman L, Petersen OB, Risor MB: Coping with worry while waiting for diagnostic results: a qualitative study of the experiences of pregnant couples following a high-risk prenatal screening result, *BMC Pregnancy Childbirth* 16(1):321, 2016.

Lou S, Frumer M, Schlutter MM, Petersen OB, Vogel I, Nielsen CP: Experiences and expectations in the first trimester of pregnancy: a qualitative study, *Health Expect* 20(6):1320–1329, 2017.

Lou S, Petersen OB, Jorgensen FS, Lund ICB, Kjaergaard S, Danish Cytogenetic Central Registry Study Group, et al: National screening guidelines and developments in prenatal diagnoses and live births of Down syndrome in 1973–2016 in Denmark, *Acta Obstet Gynecol Scand* 97(2) 195–203, 2018.

Lund ICB, Petersen OB, Becher NH, Lildballe DL, Jorgensen FS, Ambye L, et al: National data on the early clinical use of non-invasive prenatal testing in public and private healthcare in Denmark 2013–2017, *Acta Obstet Gynecol Scand* 100(5):884–892. doi:10.1111/aogs.14052. 33230826, 2021.

Malcus P, Conner P: Prenatal first trimester fetal diagnosis in Sweden today and in the future, *Lakartidningen* 114:2017.

Miltoft CB, Rode L, Tabor A: Positive view and increased likely uptake of follow-up testing with analysis of cell-free fetal DNA as alternative to invasive testing among Danish pregnant women, *Acta Obstet Gynecol Scand* 97(5):577–586, 2018.

National Statistics: *Abortion statistics for England and Wales: 2018*, 2019, National Statistics. https://www.gov.uk/government/statistics/abortion-statistics-for-england-and-wales-2018. (Accessed February 9, 2020).

Neyt M, Hulstaert F, Gyselaers W: Introducing the non-invasive prenatal test for trisomy 21 in Belgium: a cost-consequences analysis, *BMJ Open* 4(11):e005922, 2014.

Nguyen-Nielsen M, Svensson E, Vogel I, Ehrenstein V, Sunde L: Existing data sources for clinical epidemiology: Danish registries for studies of medical genetic diseases, *Clin Epidemiol* 5:249–262, 2013.

NHS Maternity Statistics: *England 2018–19*, 2019. https://digital.nhs.uk/data-and-information/publications/statistical/nhs-maternity-statistics/2018-19. (Accessed February 09, 2020).

NHS screening, 2018. https://www.nhs.uk/conditions/nhs-screening/. (Accessed February 09, 2020).

Oepkes D, Page-Christiaens GC, Bax CJ, et al: Trial by Dutch laboratories for evaluation of non-invasive prenatal testing. Part I-clinical impact, *Prenat Diagn* 36(12):1083–1090, 2016.

Petersen OB, Vogel I, Ekelund C, Hyett J, Tabor A, Danish Fetal Medicine Study Group, et al: Potential diagnostic consequences of applying non-invasive prenatal testing: population-based study from a country with existing first-trimester screening, *Ultrasound Obstet Gynecol* 43(3):265–271, 2014.

Petersson K, Lindkvist M, Persson M, Conner P, Ahman A, Mogren I: Prenatal diagnosis in Sweden 2011 to 2013-a register-based study, *BMC Pregnancy Childbirth* 16(1):365, 2016.

RKKP: *Dansk Føtalmedicinsk Database (FØTO-databasen). National årsrapport 2008–10*, 2011, RKKP.

RKKP: *Tidlig Graviditet og Abort – TiGrAb. Årsrapport 2018*, 2019, RKKP.

RKKP: *Dansk Føtalmedicinsk Database (FØTO-databasen). National årsrapport 2018*, Aarhus, 2020, RKKP.

Sas A: *Number of legal abortions in Poland 1994–2019*, 2020. https://www.statista.com/statistics/1111313/poland-number-of-legal-abortions-1994-2018/.

Schwennesen N, Koch L: Representing and intervening: 'doing' good care in first trimester prenatal knowledge production and decision-making, *Sociol Health Illn* 34(2):283–298, 2012.

Statista: *Number of abortions performed in Belgium from 2006 to 2017*, 2020. https://www.statista.com/statistics/1181758/number-of-abortions-belgium/.

Sundhedsstyrelsen: *Retningslinjer for fosterdiagnostik: prænatal information, risikovurdering, rådgivning og diagnostik*, 2004, Sundhedsstyrelsen. https://www.sst.dk/da/Udgivelser/2020/Retningslinjer-for-fosterdiagnostik

Sverige_Socialstyrelse: *Statistik om aborter 2018*, 2018. https://www.socialstyrelsen.se/globalassets/sharepoint-dokument/artikelkatalog/statistik/2019-5-18.pdf
Sveriges_Riksdag: *Socialutskottets betänkande 1989/90:SoU4: Abortlagen mm*, 1990, Sveriges Riksdag.
Szczurek-Żelazko J. On the standard of organizational perinatal care - Poland (W sprawie standardu organizacyjnego opieki okołoporodowej). 2018 [cited 2021; Available from: https://isap.sejm.gov.pl/isap.nsf/download.xsp/WDU20180001756/O/D20181756.pdf
Tabor A, Philip J, Madsen M, Bang J, Obel EB, Norgaard-Pedersen B: Randomised controlled trial of genetic amniocentesis in 4606 low-risk women, *Lancet* 1(8493):1287–1293, 1986.
THL_Finland: *Abortions in Finland 1987–2019*, 2020. https://sampo.thl.fi/pivot/prod/en/ab/kokomaa/fact_ab_kokomaa.
UK.Legislation, editor: *UK abortion act 1967*. London, 1967, UK Government.
van der Meij KRM, Sistermans EA, Macville MVE, Stevens SJC, Bax CJ, Bekker MN, et al: TRIDENT-2: National implementation of genome-wide non-invasive prenatal testing as a first-tier screening test in the Netherlands, *Am J Hum Genet* 105(6):1091–1101, 2019.
Vestergaard EM, Christensen R, Petersen OB, Vogel I: Prenatal diagnosis: array comparative genomic hybridization in fetuses with abnormal sonographic findings, *Acta Obstet Gynecol Scand* 92(7):762–768, 2013.
Vogel I, Petersen OB, Christensen R, Hyett J, Lou S, Vestergaard EM: Chromosomal microarray as primary diagnostic genomic tool for pregnancies at increased risk within a population-based combined first-trimester screening program, *Ultrasound Obstet Gynecol* 51(4):480–486, 2018.
Wapner RJ, Martin CL, Levy B, Ballif BC, Eng CM, Zachary JM, et al: Chromosomal microarray vs karyotyping for prenatal diagnosis, *N Engl J Med* 367(23):2175–2184, 2012.
Wikipedia: *Abortion in Poland*, 2020. https://en.wikipedia.org/wiki/Abortion_in_Poland.
Wulff CB, Gerds TA, Rode L, Ekelund CK, Petersen OB, Tabor A, et al: Risk of fetal loss associated with invasive testing following combined first-trimester screening for Down syndrome: a national cohort of 147,987 singleton pregnancies, *Ultrasound Obstet Gynecol* 47(1):38–44, 2016.
Zhang S: The last children of Down syndrome, *Atlantic* :2020.

CHAPTER 2

From the consulting room: personal narratives of giving prenatal information about fetal anomalies

Hajo I.J. Wildschut[1], Hilmar H. Bijma[2] and Maarten F.C.M. Knapen[2]
[1]Wildschut Consultancy, Enkhuizen, The Netherlands
[2]Department of Obstetrics and Gynaecology, Division of Obstetrics and Prenatal Medicine, Erasmus MC, Rotterdam, The Netherlands

Clinical narratives

I (HW) would like to introduce you to Mrs. Liesbeth Noomen and her husband Niels, who visited the out-patient clinic of my university hospital in Rotterdam, the Netherlands.[1] (narrative 1). They were referred by the community midwife following an abnormal 20-week fetal-anomaly scan. Prior to their visit, I received a full report from the regional ultrasound center with details of the findings. Normally, within a few days after an abnormal 20-week fetal-anomaly scan, women are scheduled for a detailed diagnostic evaluation that includes an expert fetal ultrasound examination and subsequent counseling in our tertiary center.[2] Liesbeth just had her expert fetal ultrasound examination at our university clinic. The sonographer briefed me on the details of her findings. She told me that the fetal anomaly was confined to the fetal brain. Liesbeth's unborn child was diagnosed with an isolated severe ventriculomegaly: the diameter of the lateral ventricle was 19.5 mm (normally less than 10 mm). No associated fetal structural anomalies were detected.

When Liesbeth and Niels entered my room I tried to make them feel comfortable. I gave them time to let them get aquatinted to where they were and to let them take in the meaning of the recent ultrasound findings. I encouraged them to express their feelings about their current situation. Liesbeth told me she was extremely stressed.

[1]These clinical narratives were published earlier (Van de Vathorst et al., 2008); the names of the patients are fictional.

[2]In the Netherlands a distinction is made between the fetal-anomaly scan (structureel echoscopisch onderzoek) conducted at 18-to 20-week gestation for universal fetal anomaly screening and an expert ultrasound examination [geavanceerd ultrageluidonderzoek (GUO)] for women with known risk factors, for example, diabetes, or at those with risk factors arising during pregnancy, for example, suspected fetal anomaly (GUO 2). An expert ultrasound examination is typically conducted in tertiary centers, that is, a university medical center or its satellite centers.

Prenatal Genetic Counseling
DOI: https://doi.org/10.1016/B978-0-12-823329-0.00015-5

"It is so unreal. Up to last week everything seemed okay and then, out of the blue, we are faced with this terrible news." "What to do?"

The moment I had the impression that Liesbeth and Niels felt more or less at ease, I mentioned cautiously the various management options and potential policy implications, thereby taking continuously into account whether the given information had been understood.[3] Since an isolated ventriculomegaly is associated with fetal chromosomal abnormalities, I discussed with Liesbeth and her husband the option of genetic testing. Following my explanation of the various management options, Liesbeth and Niels decided to have an amniocentesis. I conducted this procedure the same day. Before they left my clinic I gave them some practical advice. After they had left, I called her community midwife to inform her about Liesbeth's and Niels' consultation and the subsequent plan of action.

Two weeks later I saw Liesbeth and Niels again. The day before the clinical geneticist told them by telephone that the amniocentesis revealed Down syndrome (trisomy 21). At that time Liesbeth was 22-weeks pregnant. Following extensive counseling by myself and the clinical geneticist, the couple opted for pregnancy termination. We told them that their seemingly well-informed decision had our full support. Pregnancy termination was subsequently conducted at a gestational age of almost 23 weeks.[4]

My next patient[5] is Clara Pelt and her husband Mike (narrative 2). When I met them they seemed rather relaxed. At the time she was 21-weeks and 5-day pregnant. She was referred by her community midwife because of a mild intracerebral ventriculomegaly detected at the 20-weeks fetal-anomaly scan conducted the week before I saw her. She just had an expert ultrasound examination at our institution where the slight increase in the diameter of the lateral ventricles was confirmed. Apart from this finding, no other fetal structural anomalies were encountered. Clara and Mike seemed relieved despite the fact that the fetal prognosis of this finding is uncertain. I talked with them about the option of genetic testing by amniocentesis. After weighing the pros and cons of this procedure and its implications, they decided to have the amniocentesis. The same day amniocentesis was done. Chromosomal microarray showed

[3]For this purpose the teach-back method is used. This method is a way of checking understanding by asking patients to state in their own words what they need to know or do about their health condition. It is a way to confirm that you have explained things in a manner your patients understand (https://www.ahrq.gov/health-literacy/quality-resources/tools/literacy-toolkit/healthlittoolkit2-tool5.html).

[4]In the Netherlands, the law permits termination of pregnancy until a gestational age of 24 weeks under certain conditions (NVOG, 2015). Termination of pregnancy prior to 24-week gestational age is only considered at explicit parental request. Termination of pregnancy may not be a realistic option in other countries because of more restrictive legislation on abortion.

[5]Throughout the text the terms "patient," "parent," and "couple" are used interchangeably and refer to a pregnant woman (together with her life partner where indicated).

normal results. Blood tests for congenital fetal infections, including syphilis, cytomegalovirus, toxoplasma, rubella, and parvovirus infection, were negative.

During the follow-up visit at a gestational age of 23 weeks and 5 days, a slight increase in the diameter of the lateral ventricles was noted (diameter 13.7 mm). I informed them about the potential implications of the suspected anomaly in terms of future health and well-being of their baby. I also tried to explain to them the uncertainty about the natural course and subsequent prognosis of this finding for future health and well-being. I informed them about the management options, ranging from wait-and-see to pregnancy termination. Clara and Mike were quite certain what they wanted: "We have a positive attitude towards life." "Life is full of risks; we accept these as they come." "Moreover, our child is always welcome, even if its prognosis is poor." The couple decided to continue the pregnancy. I told them that we fully supported their decision since it corroborated with their own views, values, and feelings. I advised further diagnostic evaluation and monitoring. I scheduled the couple for discussion at the upcoming multidisciplinary perinatal team meeting.

Clinical dilemmas

Nowadays, prenatal screening tests are offered to every pregnant woman, at least in high resource countries (Edwards et al., 2018; Kater-Kuipers et al., 2018; Kenkhuis et al., 2018). Apart from the joy of watching their child on the ultrasound screen, the pregnant woman experiences prenatal screening as a tool to have "normality" confirmed. In fact, most pregnant women undergo prenatal screening tests to be reassured with in the back of their minds the constant concern "what if things are wrong?" (Scamell, 2011). While normal findings at ultrasound examination have strong beneficial psychological effects on the pregnant woman and her partner in terms of reassurance and parental bonding, the couple is often ill-prepared for bad news about the health of their unborn child in the case of untoward findings (Bijma et al., 2008). This news often comes out of the blue, where pregnancy, initially being considered a natural social phenomenon, turns into a very distressing medical problem. The diagnosis of a major fetal anomaly often evokes strong emotional reactions among the couple such as anxiety, disbelief, loneliness, anger, despair, prostration, grief, and guilt (Bijma et al., 2008; Lafarge et al., 2014). In fact, major fetal anomalies are typically experienced by the prospective parents as a devastating loss of a normal pregnancy (Lafarge et al., 2014). Under these circumstances, it is often very difficult for the expectant parents to grasp the nature of the condition and its possible consequences for future daily life, for themselves, their child, and their family. In fact, the couple is confronted with at least two potentially conflicting emotional values, that is, the well-being of their unborn child and the well-being of themselves and their family. How do they decide whose interest should prevail? The disturbing news about the unborn child is usually given in a medical setting, which is unfamiliar to the couple and often perceived as intimidating. In this context the couple often feels distraught, misunderstood, and powerless. They are inclined to follow medical advice at face value thereby ignoring their own feelings and personal values.

In contrast to Liesbeth and Niels (narrative 1), the finding of a mild intracerebral ventriculomegaly did not seem to upset Clara and her husband Mike (narrative 2). But were they well-informed, in the context of the prognostic uncertainty?[6] If an anomaly is suspected, further diagnostic evaluation in a tertiary center is warranted (Carta et al., 2018; Fox et al., 2018), as is demonstrated in both clinical narratives. Further diagnostic evaluation may include second-trimester amniocentesis for genetic testing, that is, microarray and whole exome sequencing (see Chapter 3). Second-trimester amniocentesis is associated with an excess risk of pregnancy loss of approximately 0.3% (95% CI, 0.11–0.49%) (Salomon et al., 2019).

Other potential investigations include maternal blood sampling for infections, magnetic resonance imaging (MRI) (Pilliod et al., 2018), and follow-up advanced ultrasound examinations (narrative 2). The latter is indicated to monitor the progression of the abnormal findings. Moreover, due to the ongoing development of the organ systems, some anomalies will not be apparent until later in the pregnancy, necessitating follow-up visits for further evaluation of the fetal condition (Edwards et al., 2018). The prospects for the future health of Clara and Mike's child are uncertain.

Management options when a fetal anomaly is detected

Major fetal structural anomalies involve management decisions about whether to apply full life-sustaining treatment even in the light of an adverse fetal prognosis in terms of survival and quality of life or to prevent suffering by taking an end-of-life decision. In Rotterdam, following the ultrasound diagnosis of a major fetal anomaly, a selection of affected women will be discussed in a multidisciplinary perinatal team.[7] Such teams have been set up in many tertiary care centers, including the Erasmus MC in Rotterdam, the Netherlands (Bijma et al., 2004a) (Box 2.1).

[6]In the setting of isolated ventriculomegaly of 10–12 mm, the likelihood of survival with normal neurodevelopment is >90%. With moderate ventriculomegaly (13–15 mm) the likelihood of normal neurodevelopment is 75%–93% (Fox et al., 2018).

[7]In Rotterdam (Erasmus MC) the woman whose pregnancy is complicated by a major fetal anomaly will be discussed in the multidisciplinary perinatal team if she meets one or more of the following criteria: (1) there is uncertainty about the diagnosis and/or prognosis, (2) when the infant is expected to be referred after birth for treatment at the neonatal unit of the Erasmus MC, and (3) when there is a parental request for pregnancy termination beyond 24 weeks of gestation, which is the upper legal limit for pregnancy termination in the Netherlands (Bijma et al., 2004a; Bijma, 2007).

BOX 2.1 Multidisciplinary team in Erasmus MC, Rotterdam, the Netherlands.

The multidisciplinary team

The multidisciplinary perinatal team in Rotterdam typically consists of medical (sub)specialists involved in perinatal care, including expert ultrasonographers; fetal and maternal medicine specialists; neonatologists; pediatric subspecialists in the fields of surgery, intensive care, cardiology, neurology, nephrology, and urology; clinical geneticists; and a variable number of physicians with different levels of expertise and experience. The multidisciplinary perinatal team meets weekly and makes decisions on obstetric and neonatal management as well as on the intended place of delivery (i.e., whether or not the delivery should take place in a tertiary care center). The Rotterdam team shares many characteristics with multidisciplinary perinatal teams in other tertiary care centers in the Netherlands and elsewhere. Differences may exist in the participation of other health-care professionals, such as social workers, midwives, nurses and general practitioners, the openness of the debate on end-of-life decisions, and the legal regulations. These factors may influence the process of decision-making in other centers.

The eligible patient is discussed by the multidisciplinary perinatal team to determine the recommended option for perinatal management. The condition of the fetus is discussed in terms of the likelihood of survival and quality of life. The multidisciplinary team sessions serve several goals, that is, the collection and exchange of all relevant information of the maternal and fetal clinical condition, reaching consensus on management decisions, thereby attuning obstetric and neonatal management decisions, and sharing the moral and medical burden of responsibility for such difficult decisions (Bijma, 2007). End-of-life decisions may be taken when, according to prevailing medical insight, the infant will not have a chance of survival or has a very poor prognosis in terms of quality of life while it is expected that a targeted treatment will not be beneficial (Box 2.2). The team strives for consensus.

BOX 2.2 Criteria for end-of-life decisions derived from the guideline of the Dutch Society of Pediatrics (Nederlandse Vereniging voor Kindergeneeskunde, 2000).

Not standing a chance: according to prevailing medical insight, the newborn infant will have no chance of survival, that is, will die immediately or within several days or months after birth or has a very poor prognosis in terms of quality of life.

Treatment is futile: apart from alleviation of suffering, fetal or neonatal treatment is deemed to be ineffective in terms of improving survival or quality of life.

In cases where maximal care is considered appropriate, the team decides to apply tailored perinatal care that typically involves a policy of referral of the pregnant woman to an obstetric unit with adequate neonatal facilities for targeted treatment, where indicated (Box 2.3). This is also true for fetal conditions where there is controversy within the team about the appropriate management (Bijma, 2007). Such

BOX 2.3 Management options for women whose pregnancy is complicated by a major fetal anomaly.

Tailored perinatal care

Timely referral of the prospective mother with targeted treatment of the newborn infant in a facility with an adequate level of obstetric and neonatal care.

Noninterventional perinatal management

Noninterventional (syn: nonaggressive) perinatal management refers to obstetric and neonatal management (e.g., striving for natural childbirth, no cesarean section for fetal distress, no active neonatal life support) in which interventions needed to sustain fetal or neonatal life are forgone because of the poor prognosis of the fetus (Bijma et al., 2008; Wool, 2011). Chervenak and McCullough (1989) first reported the noninterventional obstetric management approach in 13 women where such an approach was adopted. They regarded noninterventional obstetric management as permissible, and even preferable, when there is certainty of death or absence of cognitive developmental capacity as an outcome of the congenital anomaly. They argued that in such cases the fetus does not benefit from obstetric intervention, while such intervention may harm the pregnant woman and interfere with her autonomy (Bijma et al., 2008). Noninterventional obstetric management may be a good alternative where (late) pregnancy termination is not a (legal) option.

Comfort care

Comfort care (syn: perinatal palliative care) is intended to meet the needs of the woman (and her partner) who choose to continue a pregnancy complicated by the detection of major fetal anomaly with a poor (i.e., life-limiting) prognosis (Royal College of Obstetricians and Gynaecologists (RCOG), 2010; Rusalen et al., 2019). A multidisciplinary coordinated approach provides the couple with comprehensive, holistic support (Rusalen et al., 2019; Wool, 2013). In this context, support should be nondirective, nonjudgmental, and compassionate (Royal College of Obstetricians and Gynaecologists (RCOG), 2010; Lafarge et al., 2017).

Termination of pregnancy

Termination of pregnancy is a management option in which the pregnancy is typically terminated in the second trimester of pregnancy with the explicit intention of hastening fetal death. The first termination of pregnancy after ultrasound diagnosis of fetal anomaly was reported by Campbell et al. (1972). Their report concerned a fetus with anencephaly.

In the second-trimester pregnancy, termination of pregnancy is normally done by medical induction of labor (Wildschut et al., 2011). In advanced gestations, that is, beyond the gestational age of potential fetal viability (i.e., 21–24 weeks), medical termination of pregnancy may be preceded by feticide with intracardiac potassium chloride (KCl) (Royal College

(Continued)

BOX 2.3 (Continued)

of Obstetricians and Gynaecologists (RCOG), 2010; Statham et al., 2006) to assure that the infant is not born alive. In some countries, surgical termination of pregnancy by cervical dilatation and evacuation of the fetus and placenta (D&E) is an alternative option (Royal College of Obstetricians and Gynaecologists (RCOG), 2010; Kelly et al., 2010; Kerns et al., 2018; Lowenstein et al., 2006; Springer et al., 2018). The RCOG only recommends D&E when performed by specialist practitioners with access to the necessary instruments and who have a sufficiently large caseload to maintain their skills (Royal College of Obstetricians and Gynaecologists (RCOG), 2010; Fay et al., 2016).

decisions are typically motivated by the "negative" argument that there is no reason to deviate from standard treatment, that is, treatment aimed at keeping the fetus alive and in the most optimal condition (Bijma, 2007). The appropriateness of interventions aimed at sustaining life may be questioned in the case of a very poor fetal prognosis in terms of life expectancy and quality of life. For this reason, nonintervention policy, including noninterventional obstetric management, comfort care, withdrawal of life-saving neonatal care, or second-trimester pregnancy termination, may be considered appropriate management options (Bijma et al., 2008; Gibson et al., 2018) (Box 2.3).

Which fetal conditions qualify for the latter management options? That is a matter of controversy. In fact, there is no standard objective measure for the quantification of the severity of the fetal condition (Bijma et al., 2004a). Apart from the nonavailability of effective treatment, the "severity" of the condition of the unborn infant is determined by the diverging judgments and attitudes of those who are involved in the care of the pregnant woman and her child. This is compounded by the perspective and attitudes of the would-be parents experiencing ambivalent and strong emotional feelings when faced with end-of-life decisions (Bijma et al., 2008; Lafarge et al., 2014; Wool, 2011). In this context, birth plans could be beneficial and may provide a greater sense of control for parents (Hidalgo-Lopezosa et al., 2013; Cortezzo et al., 2019).

The concept of positive health

In 2011 a new dynamic concept of health was introduced: "health as the ability to adapt and to self-manage in the face of social, physical and emotional challenges" (Huber et al., 2011). This new concept focuses on the six dimensions of health (building blocks), as perceived by patients (Huber et al., 2016) (Box 2.4).

The positive elements of the new health concept emphasize that a person is more than his/her illness and still has a (large) potential for being healthy. It focuses on a person's strength rather than his/her weakness, it refers to self-management and individual responsibility, and it makes the relationship between patient and

BOX 2.4 The six dimensions (building blocks) of a new "holistic" concept of health (Huber et al., 2011, 2016).

The six dimensions of "positive health" are as follows:

1. bodily functions
2. mental functions and perception
3. spiritual/existential assets
4. quality of life/well-being
5. social and societal participation
6. daily functioning

health-care professional more balanced (Huber et al., 2011, 2016). The six-dimensional framework of this new concept of health could be used as an opportunity to elucidate the various components of the—dynamic—coping mechanisms of couples facing major fetal anomaly. This may facilitate counseling and subsequent shared decision-making. For instance, when the prospective parents are faced with the untoward news of their child having a major anomaly, they may prefer to think in terms of what their child may accomplish given its handicap rather than in terms of what their child will not be able to accomplish (narrative 2). This notion could be reassuring. The concept of "positive health," however, has not yet been validated among pregnant women and their partners. A potential negative element about this approach is that the couple's own personal considerations should be balanced against the future health and well-being of their unborn child (Korenromp et al., 2007). In fact, they are considered the spokespersons of their child and are personal responsible for the care and well-being of their future child. The interests of the child, however, do not necessarily corroborate with those of the couple. This conflict of interests probably plays a key role in parental decision-making about the management options, that is, whether or not to continue the pregnancy. Is their decision determined by their concerns about the future health and suffering of their child, by their anguish of the physical, social, and financial implications of having a "handicapped" child?

Risk communication and risk perception

From the ethos of the profession, it is imperative for the health-care professional not to miss a detectable fetal structural anomaly. Hence, during the ultrasound examination, the ultrasonographer is primarily focused on the detection of a structural anomaly of the unborn infant (Ahman et al., 2019). This requires time and concentration. Upon completion of the fetal examination, the abnormal findings should be properly interpreted and communicated by the ultrasonographer to the parents. Health-care professionals tend to convey information about their findings and consequent risks for the unborn infant in a rather medical-technical way. In fact, most of them view decision-making as a relatively rational process of first weighing and then ranking alternatives, after which the best option is chosen (Bijma et al., 2005; Lotto et al., 2017).

From the parents' perspective the graduality of chance in terms of probabilities is a difficult concept to understand and seems to be of little importance to them. The couple often has a binary perception of risk: probabilistic information is translated into two options: (1) the child will not be (severely) affected (2) the child will be severely affected and possibly die. As a result, even an unlikely outcome in terms of probability may dominate their decision-making when this outcome is associated with a very poor prognosis (Bijma et al., 2005). In fact, the process of parental decision-making should not solely be assessed on the basis of its rationality, but also on the basis of the diverse perceptions of the parents, including their intuition, beliefs and—disparate—values, which may be concealed by strong emotions and derealization, and their capacity to cope with uncertainty (Bijma et al., 2008, 2005; Lotto et al., 2017, 2018). These diverging perceptions of suspected fetal anomalies between health-care professionals and the couple underline the importance of a truly empathic approach by health-care providers.

Parental decision-making

Derealization, depersonalization, and the perception of choice

Derealization and depersonalization are psychological phenomena that may complicate counseling and subsequent shared decision-making. These phenomena typically occur following a traumatic event, such as the detection of a major fetal anomaly on ultrasound. Bijma (2007) interviewed prior to delivery 107 out of 161 eligible women whose pregnancy was complicated by a major fetal anomaly using a semistructured questionnaire. All women were interviewed by telephone. The presence of derealization was evaluated by asking them whether they had a feeling "as if it all wasn't true" (see narrative 1). The interviewer then invited women to elaborate on their answer. When women answered it was rather like feeling "as if it is happening to somebody else," this was coded as depersonalization. Women's perception of choice was evaluated by asking whether they felt they were faced with choices regarding the continuation of their pregnancy; 58 of 107 women (54%) felt they did not have a choice concerning perinatal management of their pregnancy. Reported feelings of derealization or depersonalization were present in 50% of all interviewed women. However, these phenomena were not related to their reported perception of choice (Bijma, 2007).

Framing

Tversky and Kahneman (1981) introduced the concept of framing: not only objective data are important in the decision-making, but also the reference point against which they are being weighed. In the context of being faced with a major fetal anomaly, the reference point of some parents might be the expectation their child is completely healthy while the reference point of others might be the opposite, the conviction that their child has a lethal problem. In fact, parents tend to attach different meanings to the same fetal structural anomaly. For example, in parents whose children are all healthy, the perception of the finding of an isolated fetal anomaly, such as cleft lip, might be completely different from those who have lost a child and are childless (Bijma et al., 2005).

Minimal account

Tversky and Kahneman also observed that people generally evaluate acts in terms of minimal account, that is, they consider only the direct consequences of the act. Parents sometimes wish to terminate pregnancy immediately after being informed of the diagnosis of a lethal anomaly (Bijma et al., 2005). In doing so, patients tend to consider the direct consequences of their decision only. They deny the possibility of considering the alternatives, such as the continuation of the affected pregnancy. Bijma et al. (2005) asserted that relief resulting from denying the situation is the direct consequence of their attitude. By getting the affected child "out of their system," parents believe that they evade the grief that results from bonding with the child they are about to lose. But in doing so, they fail to consider the long-term consequences of their stance. In fact, later they may regret not having made a "real" decision in which all options were given a fair chance. They may also conclude that they would have preferred to continue the pregnancy even knowing that their affected child would die at or shortly after birth. The ambivalence of feelings may be the result of doubts about the accuracy of the diagnosis, regrets of not having "known" their child longer, feelings of guilt about having actively denied their child any chance to live, delayed mourning because of denial, and feelings of sadness over the loss of their child (Bijma et al., 2005; Lafarge et al., 2014, 2017). The same is true for the parental decision not to consent to autopsy, a decision that might be influenced by the short-term impact of the negative emotions associated with this investigation and by their wish to leave the child's body intact. In hindsight, they acknowledge that the autopsy might have revealed important information that could have relieved their grief and could have provided a better insight into the risk of recurrence for future pregnancies (Bijma et al., 2005).

Domains of good care

Prenatal and/or perinatal management decisions, including end-of-life decisions, may evoke trying questions and potential misunderstanding among parents. Does such nonintervention management hurt the infant? Why is a diagnosis certain enough for making end-of-life decisions before birth but needs confirmation after birth? Fisher and Lafarge (2014) identified five domains of "good care" from the perspective of the affected women, that is, (1) being cared for in a timeframe and environment that feels right, (2) receiving the right level of care, (3) being supported by health-care professionals and organizations, (4) acknowledging women's particular circumstances, and (5) enabling women to make choices. Regarding the latter theme, women particularly valued being offered a choice of the method of termination, that is, medical or surgical (Fisher et al., 2015; Kerns et al., 2018). Other important features of good quality decision-making include consistency, transparency, discussing—in an unbiased way—medical and ethical aspects separately, and adequate counseling of the parents, including explanation that a noninterventional obstetric management does not necessarily mean that the infant will die after birth. Women value timely, clear, and unbiased information about the anomaly, the potential management options, including the termination of pregnancy and what to expect after the birth of the affected child (Lafarge et al., 2014). In this context, scenario-making can be an important tool to support parental decision-making following a fetal anomaly detected on ultrasound. By this approach the couple is invited to envision what the implications are at the level of daily life for the child, themselves, and the family when pregnancy is continued thereby taking into account the several possible outcomes, such as abnormal neurological and cognitive development. Furthermore, the couple is invited to envision what the scenario of pregnancy termination might be. In this way the impersonal medical-technical information about the fetal anomaly and subsequent risk is converted into more concrete scenarios that the couple may use to shape their thoughts, feelings, and ethical values (Cloin and Bijma, 2012).

Pitfalls

Decision-making following the detection of major fetal anomaly is challenging because of several pitfalls, which are summarized in Box 2.5.

Ethical aspects, diagnostic uncertainty, and limited time for a detailed diagnostic evaluation often complicate informed or shared decision-making (Bijma, 2007; Bijma et al., 2004b; Lafarge et al., 2014; Lotto et al., 2018).

Prerequisites for good aftercare

During aftercare, appropriate attention should be paid to parental grief, the meaning of loss, isolation, stigmatization, self-blame, acceptance, family support, psychological support, and counseling for future pregnancies, where indicated (Dekkers et al., 2019; Korenromp et al., 2005, 2007; Lafarge et al., 2013a,b, 2014).

BOX 2.5 Issues hampering rational decision-making following the detection of major structural anomalies by ultrasound examination.

The proper interpretation of a fetal anomaly detected by ultrasound is often taxing, since essential information, such as a genetic diagnosis or unnoticed structural defects, is often not available at the time of the ultrasound assessment, as is illustrated by narratives 1 and 2. This often prohibits a definitive diagnosis available at the time of the initial ultrasound examination (Gagnon and Genetics Committee, 2009). For example, the ultrasound intracerebral finding of mild-to-severe ventriculomegaly could be an isolated fetal anomaly or part of a (chromosomal or nonchromosomal) syndrome (Carta et al., 2018; den Hollander et al., 1998; Fox et al., 2018; Wildschut et al., 2011), as exemplified by narrative 2. Hence, further diagnostic evaluation by amniocentesis and/or fetal imaging techniques is often warranted to get the definite diagnosis together with a clearer picture of the prognosis of the unborn child.

Prenatal ultrasound is associated with both false-positive and false-negative findings (Rossi and Prefumo, 2017; Goncalves et al., 2016) and with potentially transient findings, for example, the fetal ventricular septal defect of the heart (Cho et al., 2017), fetal bronchial malformations (Nunes et al., 2015), and renal pyelectasis of the fetus (Srinivasan et al., 2013). The prognosis of the fetus with a potentially transient finding ranges from early death or severe cognitive disability to normal development.

Time available for detailed diagnostic evaluation may be limited as the result of national legislation precluding the option of abortion after a particular gestational age that varies from country to country.

Women having had a pregnancy termination for fetal anomalies may benefit from tailored counseling that includes dispelling misconceptions about the cause of the anomaly (Maguire et al., 2015). Outcome data on children born with specific anomalies are required to provide better information on natural history and prognosis. These data would enable a more accurate assignment of prognosis and better informed prenatal counseling in the future (Royal College of Obstetricians and Gynaecologists (RCOG), 2010). Qualitative studies indicate that women perceive a lack of aftercare, with some women making use of counseling services and support groups (Lafarge et al., 2013a,b, 2014; Ramdaney et al., 2015). In a study by Ramdaney et al. (2015), women cited family, friends, and genetic counselors as major help sources they would use and only a few (5.9%) of them intended to contact a grief counselor or psychologist/psychiatrist.

Coping with a major fetal anomaly and associated loss and handicap is a lengthy individual process. Coping styles are determined by women's personality, values, beliefs, family situation, culture, religion, and capacity for self-regulation (Lafarge et al., 2017).

In general, women adapt well to grief, but end-of-life decisions may have a long-lasting psychological impact on the lives of the parents and families (Kersting et al., 2009; Korenromp et al., 2005, 2007; Wool, 2011).

Seeing women at a follow-up appointment does not necessarily provide health-care professionals insight into the way women cope, both in the short term and in the long term. Women's partners are often an essential source of support. The same may be true for family and close friends. The need for professional support varies. A well-organized follow-up with coordinated care pathways could be helpful to enable women to make their own decisions and receive supportive care according their needs (Royal College of Obstetricians and Gynaecologists (RCOG), 2010; Lafarge et al., 2014). A child Internet-based cognitive behavioral therapy could be a feasible approach to effectively reduce posttraumatic stress, grief, depression, and improve overall mental health among women who experienced the loss of child (Kersting et al., 2013, 2011).

Recommendations

Before a prenatal screening test is performed, all women should be provided with balanced information about the purpose and potential outcomes of prenatal screening tests for the detection of fetal anomalies and should have an opportunity to discuss their options (Royal College of Obstetricians and Gynaecologists (RCOG), 2010; Ahman et al., 2019). Once the fetal anomaly is detected, women should be informed in a clear and unbiased way about the implications of the findings. These may include further diagnostic evaluation to obtain a definite diagnosis, the potential management options (including supportive noninterventional care) with the expected prognosis in terms of survival and quality of life. For this reason the Society of Obstetricians and Gynaecologists of Canada has set up a list of recommendations that are still valid today (Gagnon and Genetics Committee, 2009) (Box 2.6).

Being faced with a major fetal anomaly is for the pregnant woman and her family a devastating and often traumatic experience. In this stage of mind, she has to make informed decisions regarding the management of pregnancy, birth, and the postnatal period. These decisions should be based on nonjudgmental information that is provided by an empathetic health-care professional, following consultation of a multidisciplinary health-care team, where indicated.

Scenario-making may be an important tool to assist the couple in decision-making about the management of pregnancy that is complicated by a major fetal anomaly (Cloin and Bijma, 2012).

To facilitate the understanding of parental decision-making, the six dimensions of positive health could be assessed (Huber et al., 2011, 2016) (Box 2.4). To further improve communication between health-care professionals and prospective parents, the use of birth plans should be encouraged (Cortezzo et al., 2019; Hidalgo-Lopezosa et al., 2013). Further research into supporting tailored decision-making and

BOX 2.6 Evidence-based recommendations derived from the guidelines set by the Society of Obstetricians and Gynaecologists of Canada. These recommendations are intended for genetic counselors, midwives, nurses, and physicians who are involved in the prenatal care of women confronted with prenatally diagnosed isolated or multiple structural congenital anomalies (Gagnon and Genetics Committee, 2009).

1. When a fetal structural anomaly is identified, the pregnant woman should be offered a timely consultation with a trained genetic counselor and with a maternal-fetal medicine specialist and/or a medical geneticist. The counseling should be unbiased and respectful of the patient's choice, culture, religion, and beliefs. (III-A)
2. Patients should be informed that prenatal ultrasound at 18–20 weeks can detect major structural anomalies in approximately 60%–100% of such cases. (II-2A)[a]
3. When a fetal structural anomaly is suspected or identified, a referral to a tertiary ultrasound unit should be made as soon as possible to optimize therapeutic options. (II-2A)
4. In ongoing pregnancies with fetal structural anomalies, ultrasound examination should be repeated (at a frequency depending on the anomaly) to assess the evolution of the anomaly and attempt to detect other anomalies not previously identified, as this may influence the counseling as well as the obstetrical or perinatal management. (II-2B)
5. Once a fetal structural anomaly is identified by 2D ultrasound, other imaging techniques such as fetal echocardiography, 3D obstetrical ultrasound, ultrafast fetal MRI, and, occasionally, fetal X-ray and fetal CT scan (using a low-dose protocol) may be helpful in specific cases. (II-2A)
6. Parental imaging should be considered in specific cases, depending on the fetal anomaly identified (e.g., potential dominant inheritance). (III-A)
7. Parental blood testing and invasive prenatal testing may also be required to clarify the diagnosis of a fetus with isolated or multiple structural anomalies. (II-2A)
8. Women should receive information regarding the abnormal ultrasound findings in a clear, sympathetic, and timely fashion, and in a supportive environment that ensures privacy. Referral to the appropriate pediatric or surgical subspecialist(s) should be considered to provide the most accurate information possible concerning the anomaly or anomalies and the associated prognosis. (II-2B)
9. The couple should be informed that major or minor fetal structural anomalies, whether isolated or multiple, may be part of a genetic syndrome, sequence, or association, despite a normal genetic test result. (III-A)
10. If early or urgent postnatal management may be required, delivery at a center that can provide the appropriate neonatal care should be considered. (III-A)
11. When any congenital structural anomaly has been identified prenatally, a comprehensive newborn assessment is essential for diagnosis and counseling on the etiology, prognosis, and recurrence risk for future pregnancies, especially when the etiology has not been clearly identified prenatally. (III-A)
12. In cases of termination of pregnancy, stillbirth, or neonatal death, the health professional should encourage the performance of a complete autopsy by a perinatal or

(Continued)

BOX 2.6 (Continued)

pediatric pathologist to provide maximum information on the diagnosis and etiology of the structural fetal anomaly or anomalies. When a complete autopsy is refused, the health professional should encourage the performance of at least a partial or external autopsy (including X-rays, photographs, and MRI). (III-A)

[a] This estimate does not take into account the potential effectiveness of the 12- to 13-week scan for early diagnosis of fetal congenital anomalies in the cell-free DNA era (Kenkhuis et al., 2018).

Abbreviations II-2A: Evidence from well-designed cohort (prospective or retrospective) or case–control studies, preferably from more than one center or research group. There is good evidence to recommend the clinical preventive action (A); II-2B: evidence from well-designed cohort (prospective or retrospective) or case–control studies, preferably from more than one center or research group. There is fair evidence to recommend the clinical preventive action (B); III-A: opinions of respected authorities, based on clinical experience, descriptive studies, or reports of expert committees. There is good evidence to recommend the clinical preventive action (A).

bereavement processes of couples who are faced with a fetal anomaly is needed (Dekkers et al., 2019; Lafarge et al., 2013a,b, 2014).

Acknowledgments

We wish to thank Mrs. Jitske Wildschut and Dr. Frans Los for their critical comments on an earlier draft of this chapter.

References

Ahman A, Edvardsson K, Fagerli TA, Darj E, Holmlund S, Small R, et al: A much valued tool that also brings ethical dilemmas—a qualitative study of Norwegian midwives' experiences and views on the role of obstetric ultrasound, *BMC Pregnancy Childbirth* 19(1):33. Available from: http://doi.org/10.1186/s12884-019-2178-x, 2019.

Bijma HH: *Decision-making after ultrasound diagnosis of fetal abnormality*, Rotterdam, 2007, Erasmus University Rotterdam.

Bijma HH, Schoonderwaldt EM, van der Heide A, Wildschut HI, van der Maas PJ, Wladimiroff JW: Ultrasound diagnosis of fetal anomalies: an analysis of perinatal management of 318 consecutive pregnancies in a multidisciplinary setting, *Prenat Diagn* 24(11):890–895. Available from: https://doi.org/10.1002/pd.883, 2004a.

Bijma HH, van der Heide A, Wildschut HI: Decision-making after ultrasound diagnosis of fetal abnormality, *Reprod Health Matters* 16(31 Suppl):82–89, 2008. doi: 10.1016/S0968-8080(08)31372-X.

Bijma HH, Wildschut HI, van der Heide A, Passchier J, Wladimiroff JW, van der Maas PJ: Parental decision-making after ultrasound diagnosis of a serious foetal abnormality, *Fetal Diagn Ther* 20 (5):321–327, 2005. doi:86806 [pii] 10.1159/000086806.

Bijma HH, Wildschut HI, van der Heide A, van der Maas PJ, Wladimiroff JW: Obstetricians' agreement on fetal prognosis after ultrasound diagnosis of fetal anomalies, *Prenat Diagn* 24(9):713–718. Available from: https://doi.org/10.1002/pd.972, 2004b.

Campbell S, Johnstone FD, Holt EM, May P: Anencephaly: early ultrasonic diagnosis and active management, *Lancet* 2(7789):1226–1227, 1972. doi: 10.1016/s0140-6736(72)92273-8.

Carta S, Kaelin Agten A, Belcaro C, Bhide A: Outcome of fetuses with prenatal diagnosis of isolated severe bilateral ventriculomegaly: systematic review and meta-analysis, *Ultrasound Obstet Gynecol* 52 (2):165–173. Available from: https://doi.org/10.1002/uog.19038, 2018.

Chervenak FA, McCullough LB: Nonaggressive obstetric management. An option for some fetal anomalies during the third trimester, *JAMA* 261(23):3439–3440. Available from: https://doi.org/10.1001/jama.261.23.3439, 1989.

Cho YS, Park SE, Hong SK, Jeong NY, (sal)i EY: The natural history of fetal diagnosed isolated ventricular septal defect, *Prenat Diagn* 37(9):889–893. Available from: https://doi.org/10.1002/pd.5100, 2017.

Cloin C, Bijma H: *Moeten kiezen. Dilemma's rond de 20-weken echo*, Eindhoven, 2012, Uitgeverij Pepijn BV.

Cortezzo DME, Bowers K, Cameron Meyer M: Birth planning in uncertain or life-limiting fetal diagnoses: perspectives of physicians and parents, *J Palliat Med* 22(11):1337–1345, 2019.

Dekkers FHW, Go A, Stapersma L, Eggink AJ, Utens E: Termination of pregnancy for fetal anomalies: parents' preferences for psychosocial care, *Prenat Diagn* 39(8):575–587. Available from: https://doi.org/10.1002/pd.5464, 2019.

den Hollander NS, Vinkesteijn A, Schmitz-van Splunder P, Catsman-Berrevoets CE, Wladimiroff JW: Prenatally diagnosed fetal ventriculomegaly; prognosis and outcome, *Prenat Diagn* 18(6):557–566, 1998. 10.1002/(sici)1097-0223(199806)18:6 < 557::aid-pd303 > 3.3.co;2-v.

Edwards L, Hui L: First and second trimester screening for fetal structural anomalies, *Semin Fetal Neonatal Med* 23(2):102–111, 2018. 10.1016/j.siny.2017.11.005.

Fay V, Thomas S, Slade P: Maternal-fetal medicine specialists' experiences of conducting feticide as part of termination of pregnancy: a qualitative study, *Prenat Diagn* 36(1):92–99. Available from: https://doi.org/10.1002/pd.4720, 2016.

Fisher J, Lafarge C: Women's experience of care when undergoing termination of pregnancy for fetal anomaly in England, *J Reprod Infant Psychol* 33:69–87, 2014.

Fisher J, Lohr PA, Lafarge C, Robson SC: Termination for fetal anomaly: are women in England given a choice of method? *J Obstet Gynaecol* 35(2):168–172. Available from: https://doi.org/10.3109/01443615.2014.940291, 2015.

Fox NS, Monteagudo A, Kuller JA, Craigo S, Society for Maternal-Fetal Medicine (SMFM), et al: Mild fetal ventriculomegaly: diagnosis, evaluation, and management, *Am J Obstet Gynecol* 219(1):B2–B9, 2018. S0002-9378(18)30336-3 [pii]10.1016/j.ajog.2018.04.039.

Gagnon A, Genetics Committee: Evaluation of prenatally diagnosed structural congenital anomalies, *J Obstet Gynaecol Can* 31(9):875–881, 2009. 10.1016/S1701-2163(16)34307-9.

Gibson K, Hofmeyer A, Warland J: Nurses providing end-of-life care for infants and their families in the NICU: a review of the literature, *Adv Neonatal Care* 18(6):471–479, 2018. 10.1097/ANC.0000000000000533.

Goncalves LF, Lee W, Mody S, Shetty A, Sangi-Haghpeykar H, Romero R: Diagnostic accuracy of ultrasonography and magnetic resonance imaging for the detection of fetal anomalies: a blinded case-control study, *Ultrasound Obstet Gynecol* 48(2):185–192, 2016. 10.1002/uog.15774 GOV.UK. Department of Health and Social Care. Abortion statistics in England and Wales: 2018. Statistics and commentary on abortions carried out in 2018. (2020, 01–03–2020).

Hidalgo-Lopezosa P, Rodriguez-Borrego MA, Munoz-Villanueva MC: Are birth plans associated with improved maternal or neonatal outcomes? *MCN Am J Matern Child Nurs* 38(3):150–156, 2013. 10.1097/NMC.0b013e31827ea97f 00005721-201305000-00005 [pii].

Huber M, Knottnerus JA, Green L, van der Horst H, Jadad AR, Kromhout D, et al: How should we define health? *BMJ* 343:d4163. Available from: https://doi.org/10.1136/bmj.d4163, 2011.

Huber M, van Vliet M, Giezenberg M, Winkens B, Heerkens Y, Dagnelie PC, et al: Towards a 'patient-centred' operationalisation of the new dynamic concept of health: a mixed methods study, *BMJ Open* 6(1):e010091, 2016. 10.1136/bmjopen-2015-010091.

Kater-Kuipers A, Bunnik EM, de Beaufort ID, Galjaard RJH: Limits to the scope of non-invasive prenatal testing (NIPT): an analysis of the international ethical framework for prenatal screening and an interview study with Dutch professionals, *BMC Pregnancy Childbirth* 18(1):409, 2018. 10.1186/s12884-018-2050-4.

Kelly T, Suddes J, Howel D, Hewison J, Robson S: Comparing medical versus surgical termination of pregnancy at 13–20 weeks of gestation: a randomised controlled trial, *BJOG* 117(12):1512–1520. Available from: https://doi.org/10.1111/j.1471-0528.2010.02712.x, 2010.

Kenkhuis MJA, Bakker M, Bardi F, Fontanella F, Bakker MK, Fleurke-Rozema JH, et al: Effectiveness of 12–13-week scan for early diagnosis of fetal congenital anomalies in the cell-free DNA era, *Ultrasound Obstet Gynecol* 51(4):463–469. Available from: https://doi.org/10.1002/uog.17487, 2018.

Kerns JL, Light A, Dalton V, McNamara B, Steinauer J, Kuppermann M: Decision satisfaction among women choosing a method of pregnancy termination in the setting of fetal anomalies and other pregnancy complications: a qualitative study, *Patient Educ Couns* 101(10):1859–1864, 2018. S0738-3991 (18)30310-0 [pii]10.1016/j.pec.2018.06.012.

Kersting A, Dolemeyer R, Steinig J, Walter F, Kroker K, Baust K, et al: Brief Internet-based intervention reduces posttraumatic stress and prolonged grief in parents after the loss of a child during pregnancy: a randomized controlled trial, *Psychother Psychosom* 82(6):372–381, 2013. 000348713 [pii]10.1159/000348713.

Kersting A, Kroker K, Schlicht S, Baust K, Wagner B: Efficacy of cognitive behavioral internet-based therapy in parents after the loss of a child during pregnancy: pilot data from a randomized controlled trial, *Arch Womens Ment Health* 14(6):465–477. Available from: https://doi.org/10.1007/s00737-011-0240-4, 2011.

Kersting A, Kroker K, Steinhard J, Hoernig-Franz I, Wesselmann U, Luedorff K, Suslow T: Psychological impact on women after second and third trimester termination of pregnancy due to fetal anomalies versus women after preterm birth—a 14-month follow up study, *Arch Womens Ment Health* 12(4):193–201. Available from: https://doi.org/10.1007/s00737-009-0063-8, 2009.

Nederlandse Vereniging voor Kindergeneeskunde. *Richtlijnen ten behoeve van beslissingen rond het levenseinde in de neonatologie*, 2000, Nederlandse Vereniging voor Kindergeneeskunde.

Korenromp MJ, Christiaens GC, van den Bout J, Mulder EJ, Hunfeld JA, Bilardo CM, et al: Long-term psychological consequences of pregnancy termination for fetal abnormality: a cross-sectional study, *Prenat Diagn* 25(3):253–260. Available from: https://doi.org/10.1002/pd.1127, 2005.

Korenromp MJ, Page-Christiaens GC, van den Bout J, Mulder EJ, Hunfeld JA, Potters CM, et al: A prospective study on parental coping 4 months after termination of pregnancy for fetal anomalies, *Prenat Diagn* 27(8):709–716. Available from: https://doi.org/10.1002/pd.1763, 2007.

Lafarge C, Mitchell K, Fox P: Perinatal grief following a termination of pregnancy for foetal abnormality: the impact of coping strategies, *Prenat Diagn* 33(12):1173–1182. Available from: https://doi.org/10.1002/pd.4218, 2013a.

Lafarge C, Mitchell K, Fox P: Women's experiences of coping with pregnancy termination for fetal abnormality, *Qual Health Res* 23(7):924–936, 2013b. 10.1177/1049732313484198.

Lafarge C, Mitchell K, Fox P: Termination of pregnancy for fetal abnormality: a meta-ethnography of women's experiences, *Reprod Health Matters* 22(44):191–201, 2014. 10.1016/S0968-8080(14)44799-2.

Lafarge C, Mitchell K, Breeze AC, Fox P: Pregnancy termination for fetal abnormality: are health professionals' perceptions of women's coping congruent with women's accounts? *BMC Pregnancy Childbirth* 17(1):60, 2017. 10.1186/s12884-017-1238-3.

Lotto R, Smith LK, Armstrong N: Clinicians' perspectives of parental decision-making following diagnosis of a severe congenital anomaly: a qualitative study, *BMJ Open* 7(5):e014716, 2017. 10.1136/bmjopen-2016-014716.

Lotto R, Smith LK, Armstrong N: Diagnosis of a severe congenital anomaly: a qualitative analysis of parental decision making and the implications for healthcare encounters, *Health Expect* 21 (3):678–684. Available from: https://doi.org/10.1111/hex.12664, 2018.

Lowenstein L, Deutcsh M, Gruberg R, Solt I, Yagil Y, Nevo O, et al: Psychological distress symptoms in women undergoing medical vs. surgical termination of pregnancy, *Gen Hosp Psychiatry* 28(1):43–47, 2006. S0163-8343(05)00119-2 [pii]10.1016/j.genhosppsych.2005.07.006.

Maguire M, Light A, Kuppermann M, Dalton VK, Steinauer JE, Kerns JL: Grief after second-trimester termination for fetal anomaly: a qualitative study, *Contraception* 91(3):234–239, 2015. S0010-7824 (14)00787-2[pii]10.1016/j.contraception.2014.11.015.

Nederlandse Vereniging voor Obstetrie en Gynaecologie (NVOG): Herziene NVOG-richtlijn 'Zwangerschapsafbreking tot 24 weken'. Utrecht, NVOG, 2015. https://www.nvog.nl/wp-content/uploads/2018/02/Zwangerschapsafbreking-tot-24-weken-2.0-28-05-2015.pdf.

Nunes C, Pereira I, Araujo C, Santo SF, Carvalho RM, Melo A, et al: Fetal bronchopulmonary malformations, *J Matern Fetal Neonatal Med* 28(16):1996−2000. Available from: https://doi.org/10.3109/14767058.2014.984603, 2015.

Pilliod RA, Pettersson DR, Gibson T, Gievers L, Kim A, Sohaey R, et al: Diagnostic accuracy and clinical outcomes associated with prenatal diagnosis of fetal absent cavum septi pellucidi, *Prenat Diagn* 38 (6):395−401. Available from: https://doi.org/10.1002/pd.5247, 2018.

Ramdaney A, Hashmi SS, Monga M, Carter R, Czerwinski J: Support desired by women following termination of pregnancy for a fetal anomaly, *J Genet Couns* 24(6):952−960, 2015. 10.1007/s10897-015-9832-x10.1007/s10897-015-9832-x [pii].

Royal College of Obstetricians and Gynaecologists (RCOG): *Termination of pregnancy for fetal abnormality in England, Scotland and Wales. Report of a working party (website)*, 2010. Retrieved from https://www.rcog.org.uk/globalassets/documents/guidelines/terminationpregnancyreport18-may2010.pdf.

Rossi AC, Prefumo F: Correlation between fetal autopsy and prenatal diagnosis by ultrasound: a systematic review, *Eur J Obstet Gynecol Reprod Biol* 210:201−206, 2017. S0301-2115(16)31086-7 [pii] 10.1016/j.ejogrb.2016.12.024.

Rusalen F, Cavicchiolo ME, Lago P, Salvador S, Benini F: Perinatal palliative care: a dedicated care pathway, *BMJ Support Palliat Care* :. 10.1136/bmjspcare-2019-001849.

Salomon LJ, Sotiriadis A, Wulff CB, Odibo A, Akolekar R: Risk of miscarriage following amniocentesis or chorionic villus sampling: systematic review of literature and updated meta-analysis. Ultrasound Obstet Gynecol. Oct;54(4):442−451, 2019. Available from: https://doi.org/10.1002/uog.20353.

Scamell M: The swan effect in midwifery talk and practice: a tension between normality and the language of risk, *Sociol Health Illn* 33(7):987−1001. Available from: https://doi.org/10.1111/j.1467-9566.2011.01366.x, 2011.

Springer S, Gorczyca ME, Arzt J, Pils S, Bettelheim D, Ott J: Fetal survival in second-trimester termination of pregnancy without feticide, *Obstet Gynecol* 131(3):575−579. Available from: https://doi.org/10.1097/AOG.0000000000002503, 2018.

Srinivasan HB, Srinivasan N, Dhungel P, London R, Lampley C, Srinivasan G: Natural history of fetal renal pyelectasis, *J Matern Fetal Neonatal Med* 26(2):166−168. Available from: https://doi.org/10.3109/14767058.2012.722726, 2013.

Statham H, Solomou W, Green J: Late termination of pregnancy: law, policy and decision making in four English fetal medicine units, *BJOG* 113(12):1402−1411, 2006. BJO1144 [pii]10.1111/j.1471-0528.2006.01144.x.

Tversky A, Kahneman D: The framing of decisions and the psychology of choice, *Science* 211 (4481):453−458. Available from: https://doi.org/10.1126/science.7455683, 1981.

Van de Vathorst S, Verhagen AAE, Wildschut HIJ., Wolf H, Zeeman G.G, Lind J.: Termination of pregnancy after the 20-week ultrasonographic examination: haste and caution. Ned Tijdschr Geneeskd. 152: 2589−2591, 2008.

Wildschut H, Both MI, Medema S, Thomee E, Wildhagen MF, Kapp N: Medical methods for mid-trimester termination of pregnancy, *Cochrane Database Syst Rev* 1:CD005216. Available from: https://doi.org/10.1002/14651858.CD005216.pub2, 2011.

Wool C: Systematic review of the literature: parental outcomes after diagnosis of fetal anomaly, *Adv Neonatal Care* 11(3):182−192, 2011. 10.1097/ANC.0b013e31821bd92d00149525-201106000-00010 [pii].

Wool C: State of the science on perinatal palliative care, *J Obstet Gynecol Neonatal Nurs* 42(3):372−382, 2013. quiz E354−375. doi:10.1111/1552-6909.12034 S0884-2175(15)31280-6 [pii].

CHAPTER 3

Whole-exome sequencing: A changing landscape of prenatal counseling

Karin E.M. Diderich, Jasmijn E. Klapwijk, Marieke Joosten, Hennie T. Brüggenwirth and Malgorzata I. Srebniak
Department of Clinical Genetics, Erasmus Medical Centre, Rotterdam, The Netherlands

Introduction

The introduction of whole-exome sequencing (WES) in prenatal diagnostics results in an increase in resolution and diagnostic yield. This is, however, not the first time a new technique changes the landscape of prenatal counseling. With the introduction of array in prenatal diagnostics, there was also an increase in resolution and diagnostic yield as well as in unexpected diagnoses (UD) and variants of unknown clinical significance (VOUS). In the first part of this chapter, we will look at the lessons learned from introducing array analysis in prenatal diagnostics. In the second part of this chapter, we will describe guidelines and thoughts on indications for WES, counseling issues, and possible problematic results such as UD/incidental findings (IFs) and VOUS.

Lessons learned from introducing array analysis in prenatal diagnostics

Prenatal whole-genome cytogenetic diagnosis was for a very long time dependent on karyotyping, which requires time-consuming cell culturing and has a limited resolution (5–10 Mb) depending on optimal harvesting and chromosome staining conditions.

Nowadays, genomic microarray technology allows whole-genome testing at a higher resolution and can be applied to uncultured fetal material, allowing shorter reporting times when compared to classical cytogenetic techniques. Although genomic microarray testing is recommended for routine postnatal cytogenetics in cases of intellectual disability (ID) and/or multiple congenital anomalies (MCA) (Miller et al., 2010), as well as for prenatal diagnosis (PND) in cases of fetal ultrasound anomalies (American College of Obstetricians and Gynecologists, 2013), there is still ongoing discussion on its implementation for PND in fetuses without ultrasound anomalies

Prenatal Genetic Counseling
DOI: https://doi.org/10.1016/B978-0-12-823329-0.00001-5

(Fiorentino et al., 2013; Dugoff et al., 2016; Miny et al., 2013; Novelli et al., 2013; Donnelly et al., 2013). However, the recent ACOG guideline states that diagnostic testing (chorionic villus sampling or amniocentesis) options should be discussed and offered to all pregnant patients regardless of age or risk for chromosomal abnormality (Rose et al., 2020).

The main arguments against offering prenatal array testing in all indications are the chance of detecting:

- copy number variants (CNVs) causing well-described clinically significant anomalies not related to the initial indication (so-called UD or IFs),
- CNVs associated with a variable expressivity and heterogeneity of clinical features with a yet unquantifiable chance of an abnormal phenotype if found prenatally (susceptibility CNVs for neurodevelopmental disorders), and
- VOUS.

This raises several questions such as which outcomes of genomic arrays should be reported to pregnant couples? Should they be offered a choice on which possible array outcomes they wish to be informed? Is extensive genetic pretest counseling in every case necessary and feasible in clinical practice?

Indications

The increased resolution of chromosomal microarray (CMA) over routine karyotyping results in an increased detection rate of (submicroscopic) chromosomal abnormalities. As in many other countries, we started with the introduction of CMA in cases with ultrasound abnormalities (Srebniak et al., 2011, 2012; Gardiner et al., 2015) followed by the introduction of CMA in routine PND for all indications (Srebniak et al., 2013, 2015, 2016b; Van Opstal et al., 2015; Vanakker et al., 2014).

It can be disputed that when patients opt for an invasive diagnostic procedure because of, for example, an increased risk of Down's syndrome, microarray testing is preferable to karyotyping or rapid aneuploidy detection (for trisomies 13, 18, 21, and X,Y aneuploidy) (RAD stand-alone) to exclude/detect more unbalanced chromosome aberrations (Srebniak et al., 2013). Especially since many submicroscopic abnormalities cause ID and/or anomalies not detectable by fetal ultrasound examination (Srebniak et al., 2018). Are patients only interested in knowledge about fetal Down's syndrome or do they also want to be informed about other severe early-onset diseases of their child? The discussion on choosing broad or targeted diagnostics is not a new issue in prenatal diagnostics (Ogilvie et al., 2009). Boormans et al. (2010) showed that most pregnant women chose broader testing (karyotyping instead of RAD stand-alone) in a prospective study, and several years later, van der Steen et al. (2015) showed that if invasive testing is performed, most patients preferred higher resolution microarray testing (0.5 Mb) over microarray with 5 Mb resolution (comparable to postnatal karyotyping). In addition, van der Steen et al. (2018) showed that 62% of the women made an adequately informed choice, based on sufficient knowledge and attitude consistent with their choice of microarray resolution.

Pretest counseling

With the introduction of prenatal whole-genome array testing, it was anticipated that, compared to karyotyping, the frequency of genetic findings that are difficult to interpret and the frequency of diagnoses of late-onset (un)treatable disorders not related to the initial referral indication would be higher (Vetro et al., 2012). Therefore pretest genetic counseling was implemented as an obligatory element of microarray testing during pregnancy, which is reflected in several publications (Hillman et al., 2013; Srebniak et al., 2011; Darilek et al., 2008; Liao et al., 2014; Klugman et al., 2014; Evangelidou et al., 2013; Brady et al., 2014; Mademont-Soler et al., 2013; Schmid et al., 2013; Rooryck et al., 2013; Ganesamoorthy et al., 2013) and clinical guidelines (Silva et al., 2019; Vermeesch et al., 2012; Vanakker et al., 2014; Armour et al., 2018; Dugoff et al., 2016; Suela et al., 2017; South et al., 2013). Routine application of prenatal microarray led to increasing numbers of patients choosing prenatal array without sufficient numbers of medical doctors specialized in clinical genetics available for pretest counseling. When we introduced microarray diagnostics in the prenatal setting, pretest counseling was performed by a clinical geneticist. Now that microarray diagnostics is our standard care—as also described by Scott et al. (2013)—pretest counseling is performed by gynecologists/ultrasonographers in close collaboration with clinical geneticists and the laboratory specialists. We have a brochure with information about prenatal diagnostics and the different genetic tests. To further facilitate the pretest counseling, predetermined types of findings patients wish to be informed about need to be discussed. Both patients and medical specialists need a clear classification with illustrative examples to understand the possible outcomes that are discussed during pretest counseling. Srebniak et al. (2014) described an example:

- Pathogenic (for the proband, e.g., fetus):
  - Causative findings: pathogenic findings explaining the phenotype or matching the indication, for example, 22q11 microdeletion in a proband with a tetralogy of Fallot.
  - Unexpected diagnosis: pathogenic findings not explaining the phenotype or not matching the indication (early or late onset, treatable or not treatable), for example, a deletion in the DMD gene associated with Duchenne muscular dystrophy in a male fetus referred for PND because of an abnormal first-trimester screening.
  - Susceptibility CNVs: variants associated with neurodevelopmental disorders, but of extreme phenotypic heterogeneity and/or variable expressivity, for example, 16p12.1 deletion in a patient with developmental delay.
- VOUS.

- Carrier status of recessive diseases, for example, deletion of CFTR gene.
- IFs: abnormalities found by chance, unintentionally, in parents of probands, for example, mosaic Turner syndrome discovered on B-allelic frequency plot during quality control of the array profile of a pregnant woman referred for PND due to fetal ultrasound abnormalities.

When we started microarray analysis (in our case, single nucleotide polymorphism (SNP) array) in the case of fetal anomalies in clinical settings in 2009, we offered pregnant couples a choice regarding the extent to which they wished to be informed about predefined categories of genetic anomalies (Srebniak et al., 2011). At first the choices were to be informed about (1) only an abnormal result that explains the ultrasound abnormality, (2) results that probably have an adverse health effect in infancy and childhood, and (3) results that probably have an adverse health effect in adulthood (for fetus and possibly the parent). In a study, 11.1% of parents chose only option 1 (Srebniak et al., 2011). Later on, we abandoned these choices as we encountered cases where a severe early-onset disorder did not explain the ultrasound abnormality, and doctors would be posed for a huge dilemma if parents would have chosen option 1. An illustrative example is described by Joosten et al. (2016): a case of a fetus with a complex heart malformation in which a 5.0-Mb deletion in 15q11.2q13.1, associated with Angelman syndrome, was found. Where the parents had in mind to continue the pregnancy based on the ultrasound abnormality, they terminated the pregnancy after this additional information. In a study, we offered pregnant couples the additional choice of information on susceptibility CNVs. The majority of pregnant couples chose to be informed of any (susceptibility for) severe condition(s) of any age of onset (Riedijk et al., 2014). Our lab in Rotterdam has completely replaced conventional karyotyping by a whole-genome SNP array for all indications for prenatal invasive testing since July 2012 (Srebniak et al., 2013). Although specialists in PND have always been familiar with the phenomenon of UD, the number and nature of UD has expanded with the implementation of the SNP array when compared to conventional karyotyping (Armengol et al., 2012; Hillman et al., 2014).

As described by Riedijk et al. (2014), pretest counseling aids pregnant couples in handling the uncertainties that may accompany offering a broader scope of genetic PND using the array.

Possible problematic array findings: Unexpected diagnoses/incidental findings

Next to pathogenic causative array findings (explaining the fetal phenotype and therefore fitting the referral reason), so-called UD or IFs can also be made. We have previously proposed to use the term "UD" for fetal pathogenic findings not explaining the phenotype or not matching the indication while we proposed to use the term "IFs" for abnormalities found by chance, unintentionally, in parents of probands (Srebniak et al., 2014). However, internationally the term "IFs" is often used for unexpected pathogenic findings in the fetus as well as the parents (Klapwijk et al., 2021). Pathogenic findings that do not match the indication for testing or the fetal phenotype seen on ultrasound have always accompanied whole-genome cytogenetic testing (van Zwieten et al., 2005); however, because array testing has a higher resolution, they are now more frequently found. In a study on 1033 fetuses, an unexpected diagnosis was found in five cases (0.48%); one chromosomal abnormality (0.1%) concerned a microscopic finding, whereas four chromosomal abnormalities (0.39%) were submicroscopic findings (Srebniak et al., 2016a). Interpretation of microarray findings is also more complex in a prenatal setting than in a postnatal setting due to the fact that the prenatal phenotype is only described by ultrasound imaging and cannot be fully investigated. It is not always possible to assess whether the pathogenic CNV explains the fetal phenotype. Therefore it is not always easy to distinguish a causative array finding from an unexpected diagnosis. Subsequent (targeted) ultrasound examination, autopsy or postnatal examination, and follow-up may reveal additional abnormalities that may match the fetal genotype. Due to this limitation, it is difficult to determine the actual ratio between clearly pathogenic causative findings and UD in prenatal settings. Unexpected early-onset diagnoses of severe diseases may be considered an additional value of prenatal array testing if invasive sampling is performed.

Pathogenic unexpected array findings may be further subdivided into four subcategories: early-onset untreatable diseases, early-onset treatable diseases, late-onset untreatable diseases, and late-onset treatable diseases. There is no international consensus (even in postnatal settings) on what kind of UD in fetuses and IFs in parental samples should be reported, and caution should be taken in case of every late-onset (un) treatable disease. What should be reported seems to depend mainly on condition-specific factors such as disease severity, age of onset, and treatment availability and evidence indicating pathogenicity (Vanakker et al., 2014; Lohn et al., 2013). Armour et al. (2018) suggest to report pathogenic findings even if they seem not to match fetal phenotype and to provide extensive posttest genetic counseling. Postnatal guidelines suggesting release of all UD to patients and recommending a minimum list of genes that should be investigated have been published (Green et al., 2013). However, in prenatal and pediatric settings, severe ethical and psychosocial implications may arise if we would do so (Burke et al., 2013; Amor et al., 2020), especially because some of them may lead to future discrimination of individuals known to carry late-onset diseases (both treatable and untreatable) and therefore may violate the child's right to an open future (Stark et al., 2013; Dondorp et al., 2012; Townsend et al., 2014). The decision "to report or not to report" seems to be highly individual. The majority of genetic counselors and geneticists ($\sim$80%) believe that, after extensive pretest (or even preconception) counseling (Dondorp et al., 2012), the patients should be given a choice as to what kinds of UD are returned to them (Lohn et al., 2013). Since many genetic diseases are very rare and the risk of such a particular finding is very low, an appropriate balance between informing about all possibilities and preventing unnecessary anxiety in the future parents has to be found (Faas et al., 2012).

Joosten et al. (2016) evaluated whether UD made by prenatal array testing contributed to pregnancy management. In the described cohort a UD was found in 19/4043 (0.5%) pregnancies. This report includes a case with a 428-kb partial deletion in the *DMD* gene, associated with Duchenne muscular dystrophy, in a male fetus in a pregnancy at increased risk for Down's syndrome based on first-trimester screening. This finding was important information for the decision on pregnancy continuation. The parents chose to terminate the pregnancy. These are the types of cases, in which we and our patients perceive the SNP array to be of additional value over conventional karyotyping. Overall, the clinical usefulness of UDs in this report was assessed based on the couple's responses during posttest counseling and their decisions. In 16/19 cases the UD was helpful either for the couples in making a decision about the course of their pregnancy, for perinatal management or family genetic counseling. This adds another motive for offering whole-genome array during pregnancy in patients who wish broad testing of their fetus.

Possible problematic array findings: Susceptibility copy number variants for neurodevelopmental disorders

CNVs associated with incomplete penetrance, variable expressivity, and heterogeneity of clinical features (e.g., 16p11.2 microdeletion or 22q11 microduplication; so-called susceptibility CNVs for neurodevelopmental disorders) may complicate posttest counseling, if reported, and may complicate classification of array findings as well. Although there are guidelines (Kaminsky et al., 2011; Kearney et al., 2011) and such findings may be classified as pathogenic (Srebniak et al., 2014), some still classify such CNVs as VOUS (Brady et al., 2014; Wapner et al., 2012). The phenotypes of carriers of these CNVs seem to vary from normal to severely affected, and in the past the phenotypes of control individuals carrying these CNVs were shown to be intermediate between affected carriers and noncarrier control individuals (Stefansson et al., 2014), so the presence of an abnormal phenotype seems to be dependent on a "second hit" such as another genomic alteration (Veltman and Brunner, 2010).

Most guidelines recognize these susceptibility CNVs as potentially problematic; however, recommendations vary between papers (Klapwijk et al., 2021). Suela et al. (2017) and Vermeesch et al. (2012) suggest that these findings should be discussed during pretest counseling and the decisions on reporting should be given to patients. Vanakker et al. (2014) and Armour et al. (2018) suggest that only some of them should be reported, depending on the penetrance and fetal phenotype.

These CNVs, associated with neurodevelopmental disorders, but of extreme phenotypic heterogeneity and/or variable expressivity, are quite often found in prenatal settings. The literature data show a statistically significant higher frequency of these CNVs in fetuses with ultrasound abnormalities when compared to fetuses without ultrasound anomalies, 3.6% versus 0.8% (Wapner et al., 2012), 1.4% versus 0.55% (Scott et al., 2013), and 2.6% versus 1.4% (Srebniak et al., 2016a; Van Opstal et al., 2015), respectively.

The most problematic counseling issue is that the exact risk for developing a specific disease/phenotype, if such a finding is discovered prenatally, is still largely unquantifiable (Rosenfeld et al., 2013; Benn, 2013) and, therefore, the prognosis for the fetus is unsure. The available data are based on postnatal observations (Rosenfeld et al., 2013), and there are only a few population studies showing both early- and late-onset phenotypes in the carriers of susceptibility CNVs in comparison to population matched noncarriers (and not to laboratory control population). Olsen et al. (2018) studied the 22q11.2 locus as an example of the clinical challenge in genomics. They provided not only the true population prevalence of 22q11 microduplication but also showed the population-unbiased developmental trajectories of psychiatric disorders of both rearrangement types from 1 to 32 years of age. The authors have seen marginal differences in psychiatric disorders between 22q11 microdeletion and 22q11 microduplication carriers, which suggest that, for many of the psychiatric disorders, 22q11.2 microduplication carriers need the same careful and persistent clinical attention as 22q11 microdeletion patients. The posttest counseling is also facilitated by comparing these cohort results with the risks in the general population, to provide information on the possible course of the disease when a 22q11 microduplication is found prenatally.

Although many of the susceptibility CNVs are associated with phenotypes that cannot be diagnosed in utero, some may be prenatally actionable as they involve a structural abnormality (e.g., heart anomaly in case of 1q21.1 duplication). In such cases an additional ultrasound investigation should be offered (Brady et al., 2014). Olsen et al. (2018) underlined that the current diagnostic ICD-10 system was not developed to capture the milder clinical manifestations of neurodevelopmental disorders, so they presumed that individuals presenting milder symptoms may not be referred for clinical evaluation. In the case of an increased risk for a neurodevelopmental disorder, a medical follow-up may be advised to assure early interventions, when the first mild symptoms are noticed.

So far there are not many prenatal cohort evaluations. Govaerts et al. (2017) performed a study of 57 cases with a PND of a susceptibility CNV for neurodevelopmental disorders. Rapid and immediate consultation by a clinical geneticist experienced in PND was applied in all cases. They found that the severity of the ultrasound anomalies and not the diagnosis of risk for neurodevelopmental disorders was the most important factor contributing to the decision on pregnancy continuation. In a large majority of cases with milder or no ultrasound anomalies, the pregnancy was continued and a normal outcome at birth was observed. In addition, most patients were interested in both prenatal and postnatal actionability of susceptibility CNV and could cope with uncertainty related to the CNV. Follow-up studies on cases diagnosed with a susceptibility CNV are necessary to further facilitate posttest counseling. Muys et al. (2020) showed a significant difference in communication and personal–social development between children with a reported susceptibility CNV during pregnancy and the control population. Unfortunately, their study group was very small and more studies are needed.

Berger et al. (2005) suggest that the information of susceptibility CNVs may represent a burden to couples and offering such knowledge should be accompanied by careful pre- and posttest counseling. Concerns have been raised about the possible emotional harm indeterminate findings may cause to pregnant couples (McGillivray et al., 2012). Another study stated that uncertain test results do not enable choice, and, therefore, the burden of the anxiety and stress of uncertain test results is not justified (Bernhardt et al., 2013). Some have argued that the burden of generating susceptibility CNVs likely outweighs the possible benefits of detecting more pathogenic findings (Breman et al., 2012; Srebniak et al., 2013). Others argue that it is paternalistic to withhold any outcomes from PND (McGillivray et al., 2012). Instead of withholding outcomes, it has been suggested that an improved approach to reduce parental uncertainty is needed in counseling (Green et al., 2013; Wapner et al., 2012; Stark et al., 2013).

van der Steen et al. (2016) investigated the psychological impact of PND and disclosure of susceptibility CNVs in several parents who had received such a result. The majority of participants indicated they would like to be informed about the susceptibility CNV again. Most had no enduring worries. Participants unanimously indicated that pregnant couples should have an individualized pretest choice about susceptibility CNVs (non)disclosure. No negative psychological impact with the PND and disclosure of these CNVs on participants was observed. A key factor in mitigating parental anxiety with susceptibility CNV disclosure indeed appears to be posttest genetic counseling.

There is no internationally recognized policy to report susceptibility CNVs or not. Both classification and reporting of susceptibility CNVs might be questionable, but these are relatively frequent findings and they should be further investigated. It would be useful to do long-term follow-up of prenatally ascertained cases to provide more insight into the yet unquantifiable risk for an abnormal phenotype when a susceptibility CNV is prenatally detected in an uneventful pregnancy in an apparently healthy family as postnatal studies may be biased. The risks may be different in families with no history of neurodevelopmental disorders and in those with multiple family members with mild symptoms even if they were not referred for medical advice before.

Possible problematic array findings: Variants of unknown clinical significance

The question "what to report" becomes even more complicated when it concerns VOUS. There is no consensus in reporting VOUS (Klapwijk et al., 2021). Although some guidelines and recommendations are available (South et al., 2013; Riggs et al., 2012, 2020; Dugoff et al., 2016; Suela et al., 2017; Vanakker et al., 2014; Armour et al., 2018; Kearney et al., 2011), the interpretation of array data remains complex (de Leeuw et al., 2012), and there is a certain gray zone between potentially pathogenic VOUS and truly pathogenic findings, which is caused by limited (open access) knowledge and rarity of genetic diseases.

Several publications show that at least some VOUS are prenatally reported; however, there are different conditions for reporting. Some reported all VOUS (Ganesamoorthy et al., 2013), some suggest reporting VOUS, but only after extensive pre- and posttest counseling (Dugoff et al., 2016). Others report only VOUS they deem relevant (Brady et al., 2014) and others report VOUS after multidisciplinary decisions (Wapner et al., 2012). Some authors strongly recommend only report pathogenic array findings in prenatal settings (Srebniak et al., 2011; Berg et al., 2011; Vanakker et al., 2014; Brady et al., 2014). Reporting CNVs classified as benign or polymorphic, and reporting true VOUS, from which the causality is often overestimated (even if de novo) (Vermeesch et al., 2011), does not contribute to pregnancy management. In addition, if reported, these findings may even complicate the decision to continue or terminate the pregnancy (illustrated by the example described by Charan (2011). Bernhardt and colleagues presented some women to whom uncertain results were given and who continued to worry after delivery and regretted about having the test. Most patients did not even recall discussing uncertain variants during pretest counseling and many were blindsided by normal preliminary results (RAD and karyotyping) (Bernhardt et al., 2011, 2013). Moreover, Lohn et al. (2013) showed that some of the patients with a VOUS actually believed that a pathogenic CNV was found in their child. Therefore careful consideration whether a VOUS is potentially pathogenic and worth reporting (actionable) or not should be ideally made in a multidisciplinary setting (Wapner et al., 2012; Vanakker et al., 2014). If there is a local policy to report VOUS in the prenatal setting, patients should be informed about the possibility of finding and reporting VOUS (Silva et al., 2019). When reporting VOUS, pre- and posttest counseling is obligatory in guiding patients through the uncertainty (Silva et al., 2019).

Some CMA guidelines discourage reporting VOUS (Armour et al., 2018; Vanakker et al., 2014). To minimize the need for analyzing and reporting VOUS, Armour et al. (2018) advise not to report VOUS except for deletions of >500 kb and duplications of >1 Mb with emerging evidence for pathogenicity. Trio-analysis is advised to aid interpretation and decision-making concerning reporting

(Suela et al., 2017; Vanakker et al., 2014; Dugoff et al., 2016). Finally, Suela et al. (2017) suggest that patients should be given the choice, which results are reported. Thus there is no international consensus on reporting VOUS, which may be the reflection of cultural differences how the local society deals with both uncertainty in medicine and paternalistic medical guidelines (Meeuwesen et al., 2009).

Posttest counseling in practice

When prenatal array analysis reveals a chromosomal abnormality, this result should be discussed in a (daily) multidisciplinary meeting with laboratory specialists and clinical geneticists involved in prenatal care. The clinical geneticist then notifies the gynecologist who did the invasive procedure and informs the patient. In our settings, in this first phone call, the clinical geneticist shortly describes the chromosomal abnormality and invites the patient to the clinic for a meeting the same day or the day after. Unexpected bad news is not communicated to patients on Friday afternoon—unless further action is needed in the weekend—to avoid a stressful weekend prior to appropriate counseling. The official laboratory report is not sent to the general hospital system until the gynecologist and patient have been informed about the result to make sure that no one can accidently mention the result to the patient before the clinical geneticist had discussed the result with the patient. During the meeting in the clinic the chromosomal abnormality and the expected phenotype in the fetus are discussed. Comparison to prenatal cohorts is necessary and counseling solely based on postnatal biased reports is not recommended. The phenotypic diversity of chromosomal abnormalities is larger than we thought before. We also discuss the possibilities to continue or terminate the pregnancy. Depending on the individual emotional state of the patient, the procedure of termination and the emotional period after the termination are mentioned. In addition, the recurrence risk and options for next pregnancy can be explained if the patient is ready to think about perspectives and already has these questions. The patients are given written information and/or contact information for any further questions. An appointment with the gynecologist is made for further management of the pregnancy. Furthermore, we offer counseling by a psychologist to further deliberate the options, prepare for the termination and saying goodbye, or come to terms with the prospect of a child with a disability (see section 2).

Whole-exome sequencing in prenatal diagnostics

The incidence of congenital structural malformations is approximately 3% in pregnancies worldwide (Hillman et al., 2015). Most of these malformations are detectable during the second trimester of pregnancy, half of them as early as in the first trimester (Kenkhuis et al., 2018). There is a wide range of potential outcomes for fetuses with malformations depending on the type of malformation, whether an anomaly is isolated or not, and the potential underlying genetic etiology (Hillman et al., 2015). Congenital malformations vary from either mild (if isolated) anomalies (i.e., postaxial polydactyly, cleft lip) to (potentially) lethal, multisystem anomalies. When ultrasound anomalies are detected, prenatal cytogenetic diagnosis is routinely offered. In such pregnancies a CMA on DNA isolated from chorionic villi or amniocytes is recommended for optimal detection of chromosomal aberrations (American College of Obstetricians and Gynecologists, 2016b). It was shown that CMA improves diagnosis by up to 6.8% over conventional karyotyping, by detecting (sub)microscopic pathogenic CNVs in isolated and nonisolated anomalies (de Wit et al., 2014). Although microarray analysis enables testing with much higher resolution than conventional karyotyping, the cause of the abnormal phenotype remains unknown in $\sim$75% of the pregnant women referred due to an ultrasound anomaly (Srebniak et al., 2017). In many prenatal diagnostic centers, these pregnant women are currently offered additional genetic counseling. When the fetus shows specific features that allow targeted DNA testing, a targeted molecular test can be performed. When the targeted analysis shows normal results, the fetus may have a nonsyndromic birth defect or an undiagnosed genetic disorder that is not detectable with conventional karyotyping, CMA, or targeted DNA analysis. Although prenatal imaging (ultrasound, MRI) has dramatically improved, the clinical information obtained is still limited in comparison with postnatal phenotyping. If a pregnancy is continued and a child with congenital anomalies is born, sometimes the phenotype is evident after birth and (further) targeted genetic testing becomes feasible. Because the prenatal phenotype is limited to ultrasound findings similar to CMA, routine prenatal WES will improve prenatal diagnostic yield. The molecular characterization of a disease has fundamental implications in the clinical setting. The etiologic definition of the prenatal phenotype is useful to discuss the parents' reproductive choices (e.g., continuation or termination of pregnancy) of the current affected pregnancy. Not only reproductive autonomy is facilitated, but also this knowledge provides optimal birth management (e.g., planned birth at a university hospital, planned caesarian section) and allows specific early interventions after birth for the identified disease. Furthermore, it withdraws ineffective or potentially harmful investigations and/or treatments after birth (Borghesi et al., 2017). Parents can (often) be provided with detailed prognostic counseling useful to predict potential complications. And finally, molecular diagnosis enables recurrence risk assessment as well as prenatal or preimplantation diagnosis in future pregnancies (Borghesi et al., 2017; Tetreault et al., 2015).

Diagnostic yield

The technological innovation in prenatal diagnostics—from karyotyping to CMA and nowadays from limited targeted DNA analysis to WES—has substantially improved the diagnosis of previously undetectable genetic anomalies (Srebniak et al., 2017; Petrovski et al., 2019). Currently, some countries are introducing WES in prenatal genetics in cases of fetal malformations (ACMG Board of Directors, 2012, 2013; Monaghan et al., 2020; International Society for Prenatal Diagnosis et al., 2018), generating a vast amount of genetic information of the unborn child next to karyotyping or even CMA. Best et al. (2018) wrote a review on studies on prenatal exome sequencing and found a diagnostic yield varying between 6.2% and 80%. These were 16 cohort studies with five or more tested fetuses with an isolated structural abnormality, multiple abnormalities, or an increased nuchal translucency in the first trimester. These differences in diagnostic yield were largely due to inclusion criteria and partially due to trio versus singleton approaches to sequencing. The data suggested that the diagnostic yields will be greater in fetuses with multiple anomalies or in cases preselected by clinical geneticists. When looking at data of ongoing pregnancies, the diagnostic yield in fetuses with isolated or multiple abnormalities varied from 19% to 42% (Pangalos et al., 2016; Corsten-Janssen et al., 2020; Lefebvre et al., 2021; Chen et al., 2020; Dempsey et al., 2021; Normand et al., 2018; de Koning et al., 2019; Deden et al., 2020; Vora et al., 2020).

When testing fetuses with a single structural anomaly, the diagnostic yield may be different in a particular organ system. For instance, when looking at so far published cases with an isolated structural abnormality and normal array results, a diagnostic yield of 4.5%–11.5% was found in cases with heart abnormalities (van Nisselrooij et al., 2020; Mone et al., 2021; Li et al., 2020), a diagnostic yield of 7.3%–7.9% was found in cases with congenital anomalies of the kidney and urinary tract (Zhou et al., 2020; Lei et al., 2020), while a diagnostic yield of 15% was found in 46 cases with a (partial) corpus callosum agenesis (Heide et al., 2020). In cases with an increased nuchal translucency, the diagnostic yield seems to increase with an increasing thickness of the nuchal translucency (Xue et al., 2020; Yang et al., 2020). One case with an increased nuchal translucency and a relevant, but noncausative pathogenic variant did not have ultrasound abnormalities later in pregnancy (Yang et al., 2020). A very high diagnostic yield has been found in cases with a skeletal dysplasia with a molecular diagnosis in 75%–88.9% of cases (Yang et al., 2019; Chandler et al., 2018; Han et al., 2020; Tang et al., 2020).

Indications: Is it feasible to select fetuses for prenatal whole-exome sequencing based on the prenatal phenotype?

After discussions on the indications for microarray diagnostics (Oneda and Rauch, 2017), nowadays, whole-genome microarray analysis is recommended in all cases with structural anomalies on expert ultrasound scans (American College of Obstetricians and Gynecologists, 2016a), allowing an additional diagnostic yield as compared to karyotyping of up to 10% (Srebniak et al., 2012; Robson et al., 2017). The currently expected diagnostic yield of WES would be 8.5% in an unselected cohort and 15% in multisystem anomalies (Lord et al., 2019); therefore we suggest that prenatal WES may also be recommended in all cases of fetal structural anomalies.

The diagnostic yield is highest in fetal cases with MCA, and therefore in MCA there is a clear indication for WES as recognized by the professionals (Monaghan et al., 2020). However, in many such cases, the parents do not need additional genetic information urgently for making decisions on the pregnancy course; for example, in a fetus with short extremities, a small thorax and polydactyly the molecular diagnosis of Ellis–van Creveld syndrome does neither change the prognosis of the fetus nor perinatal management of the pregnancy. This is supported by the study of Lord et al. (2019) on the diagnostic yield of WES in a large unselected prenatal cohort, since 20/52 (38.5%) diagnoses were made in pregnancies that were terminated based solely on the ultrasound abnormalities. In our multidisciplinary team, we believe that the detection of genetic variants as a syndromic cause of fetal anomalies may be especially valuable in prenatal cases with a less severe phenotype or cases with an apparently isolated anomaly in which a molecular diagnosis has a larger impact on decision-making during pregnancy. In such cases, parents are facing difficult decisions based on an incomplete phenotype. It is well known that many serious syndromes have postnatal features that cannot be detected prenatally on routine or expert ultrasound scans, for example, ID or hypotonia. An example is the case of Lord et al. (2019) with an atrioventricular canal defect, where a pathogenic variant in the *ANKRD11* gene causing KGB syndrome (OMIM 148050) was detected. An ID in addition to a structural cardiac defect can significantly influence the prospective parents' decision on continuing or terminating the pregnancy. Petrovski et al. (2019) found a variant in the L1CAM gene in a male fetus with agenesis of the corpus callosum. In our practice, we have diagnosed several severe syndromic disorders in fetuses presenting with milder or isolated anomalies, for example, Cornelia de Lange syndrome in a fetus with mild intrauterine growth restriction and hypospadias (Diderich et al., 2019) and Warsaw Breakage syndrome in a fetus with apparently isolated intrauterine growth restriction in midpregnancy (Diderich et al., 2021). In cases of a treatable physical anomaly, the association with or absence of ID is most important for the parents in their decision to continue or terminate the pregnancy (Joosten et al., 2016). Several papers show a large clinical impact of prenatal WES in ongoing pregnancies on either decision-making potentially resulting in a termination of pregnancy or on pre- or perinatal management (Deden et al., 2020; Corsten-Janssen et al., 2020).

Based on our clinical experience, we propose to offer rapid prenatal WES testing to prospective parents who would only consider termination of pregnancy when a severe underlying genetic condition is identified and to prospective parents who continue the pregnancy regardless of the genetic anomaly. In the last case a molecular diagnosis can substantially influence perinatal management. A "routine" WES test could be offered to all parents of fetuses with multiple anomalies detected by ultrasound who decide to terminate the pregnancy, but for whom a genetic diagnosis is mainly valuable to assess the recurrence risk and the diagnostic protocol in future pregnancies. In such cases, WES results do not change the pregnancy management and, therefore, can be discussed postpartum.

Pretest counseling

Pretest counseling for prenatal WES needs to be delivered in an intelligible fashion to patients from all educational religious and cultural backgrounds. Due to these challenges, this is best performed by trained genetic professionals who have affinity with prenatal testing (Best et al., 2018; Horn and Parker, 2018; Kilby, 2021; Dugoff et al., 2016; Monaghan et al., 2020; International Society for Prenatal Diagnosis et al., 2018).

Realistic expectations about the likelihood of a diagnosis and the possibility of not obtaining a result before birth should be part of the pretest counseling and informed consent (International Society for Prenatal Diagnosis et al., 2018). It should be communicated what outcomes are to be expected, what type of results are possible, how likely they are, and which of these results will be reported (ACMG Board of Directors, 2012).

In fact, pretest counseling for prenatal WES is not that different from pretest counseling for prenatal array. The classification with illustrative examples of array results can be adapted to possible WES outcomes:

- Pathogenic (for the proband, e.g., fetus):
    - Causative findings: pathogenic findings explaining the phenotype or matching the indication, for example, Cornelia de Lange syndrome in a fetus with a diaphragmatic hernia.
    - Unexpected diagnosis: pathogenic findings not explaining the phenotype or not matching the indication (early or late onset, treatable or not treatable), for example, X-linked hemophilia in a male fetus with a cleft lip.
- VOUS.
- Carrier status of recessive diseases, for example, pathogenic variant in CFTR gene.
- IFs: abnormalities found by chance, unintentionally, in parents of probands, for example, maternal pathogenic variant in BRCA1 gene.

In addition, limitations of the test should be addressed (Hegde et al., 2015). WES and CMA may not completely cover all genes and both can uncover variants that may not be interpretable yet. It should be explained to families that not finding a causative variant for the primary indication does not mean there is no genetic cause for this primary indication (Hegde et al., 2015). Concerning unexpected diagnosis, pretest counseling should inform on possible test outcomes. Only in this way, possible unexpected results can be made expected. It is important to be aware that the expectations that are set in the pretest counseling are not only crucial for the pregnant couples but also for the clinicians who thus create the boundaries in which they feel comfortable to convey unexpected findings.

Also, it should be noted that while a normal result of WES in the case of a mild prenatal phenotype is reassuring for prospective parents, one can never exclude the presence of an underlying syndrome (Monaghan et al., 2020). After all, not all pathogenic variants can be detected by WES and some of the variants cannot be classified as (likely) pathogenic in the absence of the postnatal phenotype and functional studies. Obvious pathogenic variants that cannot be detected by WES are methylation defects. In a retrospective study on the potential diagnostic yield of WES in pregnancies complicated by fetal ultrasound anomalies, we found three cases of hypomethylation of H19 causing Silver—Russell syndrome (Diderich et al., 2021).

Perhaps in time—in line with pretest counseling for our standard care microarray diagnostics—pretest counseling can be performed by the gynecologists/ultrasonographers from the tertiary center in close collaboration with the clinical geneticists and laboratory specialists.

Pretest counseling for genomic test in practice

In many guidelines the importance of pre- and postcounseling is underlined. However, how this pre- and postcounseling is or should be done is rarely mentioned (Klapwijk et al., 2021). Here we describe our current practice.

As advised by the guidelines of the ISPD 2018 (International Society for Prenatal Diagnosis et al., 2018), we now offer WES in selected prenatal cases in the setting of a multidisciplinary team. After expert ultrasound imaging (and before or after chorion villus biopsy/amniocentesis) in a tertiary center, pregnant women/couples with fetal ultrasound abnormalities are referred to a clinical geneticist specialized in prenatal genetics. During WES pretest counseling, after discussing the differential diagnosis, we explain the basics of genetics [chromosomes, DNA, and when relevant inheritance patterns (e.g., dominant, recessive, X-linked)] and discuss the possibilities of further DNA testing. Rapid aneuploidy detection, karyotyping, and array analysis can be put on a scale of detection level. Most often simplified pretest counseling for array analysis has already been done by the gynecologist, but we refer to it and to chromosome analysis done by array. At this moment, we offer prenatal exome sequencing in a form of a broad gene panel [including 3357 genes associated with congenital abnormalities (MCA) and ID] analysis. Concerning WES, we discuss the different possible outcomes: no abnormalities found (with the currently used techniques), a (likely) pathogenic variant in a gene that fits the phenotype, a variant of unknown significance [which in general is not enclosed prenatally but might turn out to be a (likely) pathogenic variant after birth when more is known about the phenotype] and an unexpected diagnosis/IF. Our current policy is not to mention variants of unknown significance prenatally unless in exceptional cases, the multidisciplinary team deems a specific variant of unknown significance relevant for the pregnancy management or when a functional test is available (see the example of CPT2). Considering UD, we always reveal variants associated with early-onset or a treatable disorder, while we do not reveal variants associated with an untreatable late-onset disorder. Carriership is revealed in certain cases, for example, when a risk of at least 25% of passing the disease is found.

As examples of UD, we discuss a predisposition to (breast)cancer or a cardiomyopathy (which we always reveal to parents in line with our current policy), adult-onset dementia (which is not revealed to parents in line with our current policy), and carriership. After oral and written information has been provided, an informed consent form is signed by both parents. It is important to take time, show empathy, and be available for additional questions during this diagnostic process as well as after diagnosis (Hammond et al., 2021). Such an extensive pretest counseling is very time-consuming but enables informed choice by patients and assures understanding of the complexity or unexpected results. It is, therefore, appreciated by patients (Hammond et al., 2021).

Possible problematic results from whole-exome sequencing: Variants of unknown clinical significance

The disadvantage of WES is that many findings can be associated with uncertainty (e.g., VOUS) (Mellis et al., 2018). When comparing to microarray, there are much more hits, which cannot be evaluated manually. Automatic filtering results are necessary to discard polymorphic variants and to limit this uncertainty. A whole-exome analysis can be performed instead of analyses of selected gene panels, which increases the chance of a diagnosis, however, also further enhances the chance of finding problematic VOUS (Boycott et al., 2015).

Reporting variants of unknown significance (VOUS) in the prenatal setting is generally not desirable. The guidelines specifically describing prenatal settings advise reporting (likely) pathogenic variants and only some VOUS that may contribute to the abnormal fetal phenotype (International Society for Prenatal Diagnosis et al., 2018). How to handle VOUS is often not evident and depends on local practice (Matthijs et al., 2016; ACMG Board of Directors, 2012; Wallis et al., 2013). What practice or protocol is in place should be clear for the laboratory scientists, clinicians, and finally also to patients. If there is a local policy to report VOUS in the prenatal setting, patients should be clearly informed about the possibility of finding and reporting VOUS (ACMG Board of Directors, 2012; Van El et al., 2013). The consent document should include the types of variants reported (Hegde et al., 2015). A multidisciplinary committee must be available ad hoc to discuss difficult cases (International Society for Prenatal Diagnosis et al., 2018). Richards et al. (2015) state that VOUS should not be used in clinical decision-making. However, in some rare cases, reporting a VOUS can be justified. de Koning et al. (2019) describe a prenatal case in which they found an IF: a homozygous VOUS in UQCRC2 associated with a mitochondrial complex III deficiency (OMIM 615160) with severe neonatal metabolic acidosis. As they could not exclude the presence of a metabolic phenotype in the fetus and because screening for metabolic acidosis would be straightforward after birth, they discussed the variants with the parents, indicating a low likelihood of pathogenicity and recommended screening directly after birth (which was negative) (de Koning et al., 2019).

To minimize the amount of problematic VOUS, data sharing of single nucleotide variants (and CNVs) and fetal phenotypes is of great importance to extend the knowledge on variants and to improve variant classification (Hegde et al., 2015; Matthijs et al., 2016; Vears et al., 2018; MacArthur et al., 2014; ACMG Board of Directors, 2012; Wallis et al., 2013; Weiss et al., 2013). Furthermore, functional studies to clarify certain VOUS can be performed after birth when the fetal/postnatal phenotype is not yet explained and certain VOUS could potentially be relevant.

In general, our policy is not to report variants of unknown significance prenatally. However, we have encountered a few cases, in which we were faced with a difficult dilemma. In a few cases the multidisciplinary team decided to report the variant of unknown significance. We have, for example, encountered a similar case as de Koning et al. (2019). Around a gestational age of 20 weeks, enlarged, dense kidneys were seen in a small fetus (head circumference and femur length p0,2, abdominal circumference p0,7). In addition, the stomach was relatively large with a striking proximal duodenum, and there was a single umbilical artery. An amniocentesis was performed. The next day pretest WES counseling was done, including information about nondisclosure of variants of unknown significance. Array analysis showed no pathogenic variant. WES using a broad gene panel with ca. 3300 genes associated with MCA/ID was applied, and a homozygous VOUS in the CPT2 gene was found. Pathogenic variants in CPT2 are associated with carnitine palmitoyltransferase II (CPT II) deficiency (OMIM 600650), which varies in severity from a stress-induced myopathic form to a lethal neonatal form. As prenatally detectable enlarged polycystic kidneys can be seen in the lethal neonatal form, the laboratory specialist decided to discuss this VOUS in the multidisciplinary team. In this multidisciplinary setting, we decided to inform the parents about this variant, and we advised metabolic screening and consultation by a pediatrician specialized in metabolic disorders directly after birth. We explained that the VOUS could have no meaning at all or could cause CPT II deficiency. We said that the renal abnormalities could be caused by CPT II deficiency, but that it did not explain the other ultrasound abnormalities. Furthermore, if the VOUS was disease causing, we could not predict which form of CPT II deficiency the fetus would have. Concluding, it was a result with a lot of uncertainty. On enquiry the patient said that it was good that this VOUS was reported as it could be very relevant directly after birth, although it felt a little vague now as it was not useful during pregnancy.

However, she agreed one could not hold back the result until after birth. The patient decided to continue the pregnancy as in her eyes, the ultrasound abnormalities were not extremely severe and she would only consider a termination of pregnancy in case of extreme severe abnormalities. At a gestational age of 28 weeks, unfortunately additional ultrasound abnormalities were seen: severe intrauterine growth restriction, enlarged polycystic kidneys, oligohydramnios, and a hypoplastic thorax. We discussed the possibility of open exome analysis, but the patient refused this. A few weeks later, the ultrasound also showed cardiomegaly with pericardial effusion, intracerebral ventriculomegaly, and anhydramnios. Around a gestational age of 36 weeks the parents were devastated to find out that the pregnancy ended in a fetal demise. A pathology exam showed a severe growth restriction, lung hypoplasia, a disproportionate large and heavy heart with an atrial septal defect, and disproportionate large and heavy kidneys. Unfortunately, the fibroblasts culture (from a skin biopsy) failed, so metabolic analysis on fibroblasts was not possible. Metabolic analysis in amniocytes showed a clear decrease in CPTII activity in the patient's amniocytes compared to control amniocytes, while VLCAD activity was normal compared to control amniocytes showing that the patient's amniocytes have a CPTII deficiency. Kidney abnormalities, cardiomyopathy, structural heart anomalies, and ventriculomegaly have been described in CPTII deficiency (Dorum et al., 2019; Du et al., 2017; He and Li, 2020; Meir et al., 2009). The combination of cardiomyopathy with pericardial effusion, kidney abnormalities with anhydramnios, and, in particular, the severe growth restriction can explain the fetal death. Based on the ACMG criteria, the variant was reclassified to a likely pathogenic variant. Both parents are carriers, and now invasive prenatal diagnoses can be offered for the recurrence risk of 25%. As not the whole phenotype can be explained by the CPTII deficiency, further genetic analysis was offered. Full exome analysis showed no additional abnormalities.

Possible problematic results from whole-exome sequencing: Unexpected diagnoses/incidental findings

Any test result can reveal a pathogenic variant unrelated to the primary indication, but known to be associated with an abnormal phenotype (ACMG Board of Directors, 2012). We have previously proposed to use the term "UD" for fetal pathogenic findings not explaining the phenotype or not matching the indication, while we proposed to use the term "IFs" for abnormalities found by chance, unintentionally, in parents of probands (Srebniak et al., 2014). However, internationally the term "IFs" is often used for unexpected pathogenic findings in the fetus as well as the parents (Klapwijk et al., 2021). Van El et al. (2013) recommend reporting these IFs that are indicative of serious health problems and (clinically) actionable. Hegde and colleagues recommend that only known and (likely) pathogenic IFs should be included in the test report, which is in line with the recommendation of not reporting VOUS. IFs without health implications or VOUS should not be reported (Armour et al., 2018; Gardiner et al., 2015; Vears et al., 2018).

There are several guidelines and recommendation papers that advise on the management of such findings. It should be clear to the patient what kind of IFs are excluded from reporting (e.g., adult-onset conditions), which are included (Matthijs et al., 2016), and what the psychological (and potential insurance) risks are of receiving such findings (Hegde et al., 2015). Information on the chance of identification of IFs should be given (Matthijs et al., 2016) by a medical geneticist or genetic counselor during pretest counseling (ACMG Board of Directors, 2012). Matthijs et al. (2016) recommend providing multiple sources for this information in the form of written leaflets and online information. Many authors recommend that there should be a protocol or policy in place related to disclosure of IFs (ACMG Board of Directors, 2012; International Society for Prenatal Diagnosis et al., 2018; Van El et al., 2013; Matthijs et al., 2016; Hegde et al., 2015). It is supported for patients to be able to opt-in or opt-out of receiving IFs and the informed consent should clearly indicate the availability of this choice (ACMG Board of Directors, 2012; Hegde et al., 2015). Exceptions need to be discussed multidisciplinary on a case-by-case basis (ACMG Board of Directors, 2012; International Society for Prenatal Diagnosis et al., 2018). Not all recommendations are easy to employ in clinical practice, especially when having a patient with low educational status, when a counselor is doubting whether the choice is made based on sufficient knowledge and understanding. However, Vears et al. (2018) recommend to respect a patients' choice to opt-out, even if the IF may be relevant to their health.

With regards to minors a policy should be in place that takes into consideration both the child's autonomy and interests and the parents' rights and needs, which they also believe applies prenatally (Vears et al., 2018). When it concerns children (postnatally), Vears et al. (2018) recommend to (1) report IFs that have health implications for the proband and are actionable (2) discuss on a case-by-case basis whether adult-onset conditions with significant health implications (for the proband and the family) should be returned, taking into account the health risk, actionability, and the informed consent of the parents; and (3) report carrier status in line with the informed consent, due to reproductive options.

It should be noted that finding an actionable/treatable IF is not a rarity. Haer-Wigman et al. (2019) found that 2.7% of healthy individuals in an outbred population of European descent had a (likely) pathogenic variant in 1 of 59 medically actionable dominant disease genes for which the American College of Medical Genetics and Genomics recommends disclosure (Kalia et al., 2017).

In addition to treatable/not treatable, early-onset/late-onset disorders, trio-WES can also reveal unforeseen issues such as nonpaternity or parental consanguinity, but this is not unique to WES as this can also be found when performing a SNP array.

We have encountered a few cases with an unexpected diagnosis, for example, a susceptibility to cancer or cardiomyopathy. Our policy has been to report these as discussed during pretest counseling and stated in the informed consent form. Patients are counseled about the unexpected diagnosis and referred for appropriate health monitoring for the particular disease.

Our local policy is not to report late-onset, untreatable disorders. However, sometimes an unexpected diagnosis is not so unexpected when looking at the family history. Using array, we encountered a maternal deletion of 17p associated with hereditary neuropathy with liability to pressure palsies (HNPP, OMIM 162500) in a fetus of a pregnant woman who was also undergoing neurological screening because of complaints. In such a case, disclosing a late-onset, untreatable disorder can enable rapid diagnosis for the patients' current neurological problems. It is thus important to investigate the family tree as a part of pretest genetic counseling and to discuss these cases in a multidisciplinary team and to be able to make tailor-made decisions about specific cases. The clinical information on both the parents of the fetus and family can be crucial for decisions on disclosing both UD and potentially pathogenic VOUS that match the current medical problems in the fetus or family.

Posttest counseling

Posttest counseling by a clinical geneticist is of utmost importance in case of an abnormal result. When conveying abnormal results, this should be done by the multidisciplinary team, which should include laboratory specialists, clinical geneticists, and relevant clinical specialists, and needs to include the significance of the finding, details of the diagnosed condition, including the prognosis and potential therapies (Best et al., 2018; Klapwijk et al., 2021). The logistics of reporting an abnormal result is rather similar to the logistics of informing patients and a referring fetal medicine specialist on an abnormal array result, which is described in the section on posttest counseling of abnormal array results. As we experience that the situation and information load is highly influencing patients, who first have to cope with the current situation and decide on the course of the current pregnancy, it is advisable to have a follow-up meeting to discuss the details of the recurrence risk and future reproductive options.

In case of a normal result, it should be explained to prospective parents that a genetic cause or syndrome is not entirely excluded. Also after the child is born, new phenotypic information may become available urging for reevaluation of the WES data that may—together with new literature data—reveal a pathogenic finding. When the prenatal WES is more common in clinical practice, perhaps one could argue that pretest counseling, including this point, may be sufficient in case of normal results.

A changing landscape of prenatal counseling

Because of the technological advances, prenatal genetic counseling has changed considerably. In the days of karyotyping, large chromosomal aberrations that were found in fetuses with ultrasound anomalies were per definition classified as pathogenic and were predicted to cause an abnormal fetal phenotype. However, on the other hand, some results also carried uncertainty. Some fetuses carried marker chromosomes of unknown origin with uncertain clinical significance. Clinical geneticists were used to counsel on such uncertainties, and in many cases, subsequent parental studies were needed for the risk assessment. Incidental diagnoses in the form of, for example, familiar balanced chromosomal aberrations were also encountered. However, with the introduction of microarrays, the number of patients with uncertain or unexpected results was magnified due to the detection of susceptibility CNVs and in a lower percentage UD/IFs.

With the introduction of prenatal WES the chance/risk of finding an unexpected diagnosis/IF has further increased. Therefore the higher incidence of challenging counseling sessions may be experienced as a burden by clinical geneticists. While this might be a challenge for the counselor, it also conveys health benefits for patients: knowledge of carriership can be relevant for reproductive options and health monitoring can be arranged in case of treatable/actionable disorders.

Another problematic issue is the fact that due to high-throughput techniques and therefore enormous growth of data, the need of updating the knowledge and current guidelines is urgent (Klapwijk et al., 2021). Continuous education is necessary for physicians and patients to understand the changes in molecular interpretation (Liu et al., 2019). In the past a normal karyotype was very seldom assessed as abnormal after birth, and almost all cases could be explained by the differences in banding resolution or tissue-specific aberrations. Nowadays, the development of new techniques and rapid accrual of knowledge in genomic medicine made the chance of changing the patients' diagnosis after reanalysis of existing genomic data much higher, when compared to karyotyping. A VOUS discovered by microarray or WES testing can be reclassified to a (likely) pathogenic finding based on additional information after birth or new knowledge in time. New diagnoses can not only be made due to the discovery of new genes but also by reclassifying already reported variants (Liu et al., 2019). When it happens, it may cause distress to clinicians and patients. If this issue is not sufficiently addressed during pre- or posttest counseling, it may cause an uncomfortable feeling of misdiagnosis during pregnancy. Due to the complexity of the techniques that are nowadays available in clinics, awareness of the test limitations and good communication with patients became much more important than before.

Future perspectives

We anticipate that, similar to microarray, prenatal WES will be offered in all cases (with ultrasound anomalies) that undergo invasive sampling. In addition, as the costs of sequencing fall, and bioinformatic and analytic capabilities improve and rapid sequencing becomes available, we may switch from prenatal WES and microarray approaches toward whole-genome sequencing (WGS) (Best et al., 2018). Because of improved ability to detect a wider range of genomic abnormalities, including, for example, noncoding variants, this will improve diagnostic capabilities further (Best et al., 2018). It is likely that, in the future, more data will become available on the use of prenatal WGS (Kilby, 2021). WGS is also more suitable for CNV analysis, so we presume that, in the near future, the request for microarray diagnostics and WES test will be replaced by a request for whole-genome genotyping, and the results of both variant and CNV analysis will be done based on whole-genome fetal sequencing. This will reduce pretest counseling to one session, and we predict that it will be beneficial for patients who will not be blinded by the normal initial results from RAD and microarray. Posttest counseling can then concentrate on final results without the need for "in-between" counseling for multiple laboratory tests, and the whole stressful "diagnostic roller-coaster" will be simplified.

References

ACMG Board of Directors: Points to consider in the clinical application of genomic sequencing, *Genet Med* 14:759, 2012.

ACMG Board of Directors: Points to consider for informed consent for genome/exome sequencing, *Genet Med* 15:748–749, 2013.

American College of Obstetricians and Gynecologists: Committee Opinion No. 581: the use of chromosomal microarray analysis in prenatal diagnosis, *Obstet Gynecol* 122:1374–1377, 2013.

American College of Obstetricians and Gynecologists: Microarrays and next-generation sequencing technology: the use of advanced genetic diagnostic tools in obstetrics and gynecology. ACOG Committee opinion no. 682. American College of Obstetricians and Gynecologists, *Obstet Gynecol* 128:e262–e268, 2016a.

American College of Obstetricians and Gynecologists: Practice bulletin no. 162: prenatal diagnostic testing for genetic disorders, *Obstet Gynecol* 127:e108–e122, 2016b.

Amor DJ, Chitty LS, Van den Veyver IB: Current controversies in prenatal diagnosis 2: the 59 genes ACMG recommends reporting as secondary findings when sequencing postnatally should be reported when detected on fetal (and parental) sequencing, *Prenat Diagn* 40(12):1508–1514, 2020.

Armengol L, Nevado J, Serra-Juhe C, Plaja A, Mediano C, Garcia-Santiago FA, et al: Clinical utility of chromosomal microarray analysis in invasive prenatal diagnosis, *Hum Genet* 131:513–523, 2012.

Armour CM, Dougan SD, Brock J-A, Chari R, Chodirker BN, DeBie I, et al: Practice guideline: joint CCMG-SOGC recommendations for the use of chromosomal microarray analysis for prenatal diagnosis and assessment of fetal loss in Canada, *J Med Genet* 55:215–221, 2018.

Benn PA: Prenatal counseling and the detection of copy-number variants, *Genet Med* 15:316–317, 2013.

Berg JS, Khoury MJ, Evans JP: Deploying whole genome sequencing in clinical practice and public health: meeting the challenge one bin at a time, *Genet Med* 13:499–504, 2011.

Berger G, Bernhardt T, Weimer E, Peters J, Kratzsch T, Frolich L: Longitudinal study on the relationship between symptomatology of dementia and levels of subjective burden and depression among family caregivers in memory clinic patients, *J Geriatr Psychiatry Neurol* 18:119–128, 2005.

Bernhardt B, Soucier D, Hanson K, Savage M: It's a little bit of a black box: patient and provider experiences with the uncertainties of prenatal microarray testing. In *Abstract presented at ACMG annual clinical genetics meeting*, Vancouver, Canada, 2011, American College of Medical Genetics.

Bernhardt BA, Soucier D, Hanson K, Savage MS, Jackson L, Wapner RJ: Women's experiences receiving abnormal prenatal chromosomal microarray testing results, *Genet Med* 15:139–145, 2013.

Best S, Wou K, Vora N, Van der Veyver IB, Wapner R, Chitty LS: Promises, pitfalls and practicalities of prenatal whole exome sequencing, *Prenat Diagn* 38:10–19, 2018.

Boormans EM, Birnie E, Oepkes D, Boekkooi PF, Bonsel GJ, van Lith JM, et al: Individualized choice in prenatal diagnosis: the impact of karyotyping and standalone rapid aneuploidy detection on quality of life, *Prenat Diagn* 30:928–936, 2010.

Borghesi A, Mencarelli MA, Memo L, Ferrero GB, Bartuli A, Genuardi M, et al: Intersociety policy statement on the use of whole-exome sequencing in the critically ill newborn infant, *Ital J Pediatr* 43:100, 2017.

Boycott K, Hartley T, Adam S, Bernier F, Chong K, Fernandez BA, et al: The clinical application of genome-wide sequencing for monogenic diseases in Canada: position statement of the Canadian College of Medical Geneticists, *J Med Genet* 52:431–437, 2015.

Brady PD, Delle Chiaie B, Christenhusz G, Dierickx K, Van Den Bogaert K, Menten B, et al: A prospective study of the clinical utility of prenatal chromosomal microarray analysis in fetuses with ultrasound abnormalities and an exploration of a framework for reporting unclassified variants and risk factors, *Genet Med* 16:469–476, 2014.

Breman A, Pursley AN, Hixson P, Bi W, Ward P, Bacino CA, et al: Prenatal chromosomal microarray analysis in a diagnostic laboratory; experience with >1000 cases and review of the literature, *Prenat Diagn* 32:351–361, 2012.

Burke W, Matheny Antommaria AH, Bennett R, Botkin J, Clayton EW, Henderson GE, et al: Recommendations for returning genomic incidental findings? We need to talk!, *Genet Med* 15:854–859, 2013.

Chandler N, Best S, Hayward J, Faravelli F, Mansour S, Kivuva E, et al: Rapid prenatal diagnosis using targeted exome sequencing: a cohort study to assess feasibility and potential impact on prenatal counseling and pregnancy management, *Genet Med* 20:1430–1437, 2018.

Charan, P. SNP array in cases of fetal defects. In *Oral presentation at 10th world congress in fetal medicine, 26–30 June 2011*, Malta, 2011

Chen M, Chen J, Wang C, Chen F, Xie Y, Li Y, et al: Clinical application of medical exome sequencing for prenatal diagnosis of fetal structural anomalies, *Eur J Obstet Gynecol Reprod Biol* 251:119–124, 2020.

Corsten-Janssen N, Bouman K, Diphoorn JCD, Scheper AJ, Kinds R, El Mecky J, et al: A prospective study on rapid exome sequencing as a diagnostic test for multiple congenital anomalies on fetal ultrasound, *Prenat Diagn* 40(10):1300–1309, 2020.

Darilek S, Ward P, Pursley A, Plunkett K, Furman P, Magoulas P, et al: Pre-and postnatal genetic testing by array-comparative genomic hybridization: genetic counseling perspectives, *Genet Med* 10:13–18, 2008.

de Koning MA, Haak MC, Adama van Scheltema PN, Peeters-Scholte C, Koopmann TT, Nibbeling EAR, et al: From diagnostic yield to clinical impact: a pilot study on the implementation of prenatal exome sequencing in routine care, *Genet Med* 21:2303–2310, 2019.

de Leeuw N, Dijkhuizen T, Hehir-Kwa JY, Carter NP, Feuk L, Firth HV, et al: Diagnostic interpretation of array data using public databases and internet sources, *Hum Mutat* 33(6):930–940, 2012.

de Wit MC, Srebniak MI, Govaerts LC, Van Opstal D, Galjaard RJ, Go AT: Additional value of prenatal genomic array testing in fetuses with isolated structural ultrasound abnormalities and a normal karyotype: a systematic review of the literature, *Ultrasound Obstet Gynecol* 43:139–146, 2014.

Deden C, Neveling K, Zafeiropopoulou D, Gilissen C, Pfundt R, Rinne T, et al: Rapid whole exome sequencing in pregnancies to identify the underlying genetic cause in fetuses with congenital anomalies detected by ultrasound imaging, *Prenat Diagn* 40:972–983, 2020.

Dempsey E, Haworth A, Ive L, Dubis R, Savage H, Serra E, et al: A report on the impact of rapid prenatal exome sequencing on the clinical management of 52 ongoing pregnancies; a retrospective review, *BJOG* 128(6):1012–1019, 2021.

Diderich K, Joosten M, Govaerts L, Van Opstal D, Go A, Knapen M, et al: Is it feasible to select fetuses for prenatal WES based on the prenatal phenotype? *Prenat Diagn* 39:1039–1040, 2019.

Diderich KEM, Romijn K, Joosten M, L.C.P., et al: The potential diagnostic yield of whole exome sequencing in pregnancies complicated by fetal ultrasound anomalies, *Acta Obstet Gynecol Scand* 100 (6):1106–1115, 2021.

Dondorp W, Sikkema-Raddatz B, de Die-Smulders C, de Wert G: Arrays in postnatal and prenatal diagnosis: an exploration of the ethics of consent, *Hum Mutat* 33:916–922, 2012.

Donnelly JC, Wapner R, Simpson JL: Can microarray analysis reveal additional abnormalities in prenatal diagnosis? *Expert Rev Obstet Gynecol* 8:193–195, 2013.

Dorum S, Varal IG, Gorukmez O, Dogan P, Ekici A: A novel mutation leading to the lethal form of carnitine palmitoyltransferase type-2 deficiency, *J Pediatr Endocrinol Metab.* 32(7):781–783, 2019.

Du S-H, Zhang F, Yu Y-G, Chen C-X, Wang H-J, Li D-R: Sudden infant death from neonate carnitine palmitoyl transferase II deficiency, *Forensic Sci Int* 278:e41–e44, 2017.

Dugoff L, Norton ME, Kuller JA, Society for Maternal-Fetal Medicine: The use of chromosomal microarray for prenatal diagnosis, *Am J Obstet Gynecol* 215:B2–B9, 2016.

Evangelidou P, Alexandrou A, Moutafi M, Ioannides M, Antoniou P, Koumbaris G, et al: Implementation of high resolution whole genome array CGH in the prenatal clinical setting: advantages, challenges, and review of the literature, *Biomed Res Int* 2013:346762, 2013.

Faas BH, Feenstra I, Eggink AJ, Kooper AJ, Pfundt R, van Vugt JM, et al: Non-targeted whole genome 250K SNP array analysis as replacement for karyotyping in fetuses with structural ultrasound anomalies: evaluation of a one-year experience, *Prenat Diagn* 32:362–370, 2012.

Fiorentino F, Napoletano S, Caiazzo F, Sessa M, Bono S, Spizzichino L, et al: Chromosomal microarray analysis as a first-line test in pregnancies with a priori low risk for the detection of submicroscopic chromosomal abnormalities, *Eur J Hum Genet* 21:725–730, 2013.

Ganesamoorthy D, Bruno DL, McGillivray G, Norris F, White SM, Adroub S, et al: Meeting the challenge of interpreting high-resolution single nucleotide polymorphism array data in prenatal diagnosis: does increased diagnostic power outweigh the dilemma of rare variants? *BJOG* 120:594–606, 2013.

Gardiner C, Wellesley D, Kilby MD, Kerr B, Joint Committee on Genomics in Medicine: *Recommendations for the use of chromosome microarray in pregnancy*, London, The Royal College of Pathologists, PUB, 290615, 2015.

Govaerts L, Srebniak M, Diderich K, Joosten M, Riedijk S, Knapen M, et al: Prenatal diagnosis of susceptibility loci for neurodevelopmental disorders—genetic counseling and pregnancy outcome in 57 cases, *Prenat Diagn* 37:73–80, 2017.

Green RC, Berg JS, Grody WW, Kalia SS, Korf BR, Martin CL, et al: ACMG recommendations for reporting of incidental findings in clinical exome and genome sequencing, *Genet Med* 15:565–574, 2013.

Haer-Wigman L, van der Schoot V, Feenstra I, Vulto-van Silfhout AT, Gilissen C, Brunner HG, et al: 1 in 38 individuals at risk of a dominant medically actionable disease, *Eur J Hum Genet* 27:325–330, 2019.

Hammond J, Jasmijn EK, Hill M, Lou S, Ormond KE, Diderich KEM, et al: Parental experiences of uncertainty following an abnormal fetal anomaly scan: insights using Han's taxonomy of uncertainty, *Genet Couns* 30(1):198–210, 2021.

Han J, Yang YD, He Y, Liu WJ, Zhen L, Pan M, et al: Rapid prenatal diagnosis of skeletal dysplasia using medical trio exome sequencing: benefit for prenatal counseling and pregnancy management, *Prenat Diagn* 40:577–584, 2020.

He Y, Li DZ: Carnitine palmitoyltransferase II deficiency in a prenatal case with polycystic kidney disease-like phenotype, *Congenit Anom (Kyoto)* 60(4):131–132, 2020.

Hegde M, Bale S, Bayrak-Toydemir P, Gibson J, Jeng LJB, Joseph L, et al: Reporting incidental findings in genomic scale clinical sequencing—a clinical laboratory perspective: a report of the Association for Molecular Pathology, *J Mol Diagn* 17:107–117, 2015.

Heide S, Spentchian M, Valence S, Buratti J, Mach C, Lejeune E, et al: Prenatal exome sequencing in 65 fetuses with abnormality of the corpus callosum: contribution to further diagnostic delineation, *Genet Med* 22(11):1887–1891, 2020.

Hillman SC, McMullan DJ, Hall G, Togneri FS, James N, Maher EJ, et al: Prenatal chromosomal microarray use: a prospective cohort of fetuses and a systematic review and meta-analysis, *Ultrasound Obstet Gynecol* 41(6):610–620, 2013.

Hillman SC, McMullan DJ, Silcock L, Maher ER, Kilby MD: How does altering the resolution of chromosomal microarray analysis in the prenatal setting affect the rates of pathological and uncertain findings? *J Matern Fetal Neonatal Med* 27(7):649–657, 2014.

Hillman SC, Willams D, Carss KJ, McMullan DJ, Hurles ME, Kilby MD: Prenatal exome sequencing for fetuses with structural abnormalities: the next step, *Ultrasound Obstet Gynecol* 45:4–9, 2015.

Horn R, Parker M: Opening Pandora's box?: ethical issues in prenatal whole genome and exome sequencing, *Prenat Diagn* 38:20–25, 2018.

International Society for Prenatal Diagnosis, Society for Maternal and Fetal Medicine, Perinatal Quality Foundation: Joint Position Statement from the International Society for Prenatal Diagnosis (ISPD), the Society for Maternal Fetal Medicine (SMFM), and the Perinatal Quality Foundation (PQF) on the use of genome-wide sequencing for fetal diagnosis, *Prenat Diagn* 38:6–9, 2018.

Joosten M, Diderich KE, Van Opstal D, Govaerts LC, Riedijk SR, Prinsen AK, et al: Clinical experience of unexpected findings in prenatal array testing, *Biomark Med* 10:831–840, 2016.

Kalia SS, Adelman K, Bale SJ, Chung WK, Eng C, Evans JP, et al: Recommendations for reporting of secondary findings in clinical exome and genome sequencing, 2016 update (ACMG SF v2. 0): a policy statement of the American College of Medical Genetics and Genomics, *Genet Med* 19:249–255, 2017.

Kaminsky EB, Kaul V, Paschall J, Church DM, Bunke B, Kunig D, et al: An evidence-based approach to establish the functional and clinical significance of copy number variants in intellectual and developmental disabilities, *Genet Med* 13:777–784, 2011.

Kearney HM, Thorland EC, Brown KK, Quintero-Rivera F, South ST, Working Group of the American College of Medical Genetics Laboratory Quality Assurance Committee: American College of Medical Genetics standards and guidelines for interpretation and reporting of postnatal constitutional copy number variants, *Genet Med* 13:680–685, 2011.

Kenkhuis MJA, Bakker M, Bardi F, Fontanella F, Bakker MK, Fleurke-Rozema JH, et al: Effectiveness of 12–13-week scan for early diagnosis of fetal congenital anomalies in the cell-free DNA era, *Ultrasound Obstet Gynecol* 51:463–469, 2018.

Kilby MD: The role of next generation sequencing in the investigation of ultrasound identified fetal structural anomalies, *BJOG* 128(2):420–429, 2021.

Klapwijk JE, Srebniak MI, Go Attie TJI, Govaerts LCP, Lewis C, Hammond J, et al: How to deal with uncertainty in prenatal genomics: A systematic review of guidelines and policies, *Clin Genet* 100 (6):647–658, 2021.

Klugman S, Suskin B, Spencer BL, Dar P, Bajaj K, Powers J, et al: Clinical utility of chromosomal microarray analysis in prenatal diagnosis: report of first 6 months in clinical practice, *J Matern Fetal Neonatal Med* 27(13):1333–1338, 2014.

Lefebvre M, Bruel AL, Tisserant E, Bourgon N, Duffourd Y, Collardeau-Frachon S, et al: Genotype-first in a cohort of 95 fetuses with multiple congenital abnormalities: when exome sequencing reveals unexpected fetal phenotype-genotype correlations, *J Med Genet* 58(6):400–413, 2021.

Lei T-ying, Fu F, Li R, Yu Q-xia, Du K, Zhang W-wen, et al: Whole-exome sequencing in the evaluation of fetal congenital anomalies of the kidney and urinary tract detected by ultrasonography, *Prenat Diagn* 40 (10):1290–1299, 2020.

Li R, Fu F, Yu Q, Wang D, Jing X, Zhang Y, et al: Prenatal exome sequencing in fetuses with congenital heart defects, *Clin Genet* 98:215–230, 2020.

Liao C, F. Fu RL, Xie GE, Zhang YL, Li J, Li DZ: Implementation of high-resolution SNP arrays in the investigation of fetuses with ultrasound malformations: five years of clinical experience, *Clin Genet* 86 (3):264–269, 2014.

Liu P, Meng L, Normand EA, Xia F, Song X, Ghazi A, et al: Reanalysis of clinical exome sequencing data, *N Engl J Med* 380:2478–2480, 2019.

Lohn Z, Adam S, Birch P, Townsend A, Friedman J: Genetics professionals' perspectives on reporting incidental findings from clinical genome-wide sequencing, *Am J Med Genet A* 161A:542–549, 2013.

Lord J, McMullan DJ, Eberhardt RY, Rinck G, Hamilton SJ, Quinlan-Jones E, et al: Prenatal exome sequencing analysis in fetal structural anomalies detected by ultrasonography (PAGE): a cohort study, *Lancet* 393(10173):747–757, 2019.

MacArthur DG, Manolio TA, Dimmock DP, Rehm HL, Shendure J, Abecasis GR, et al: Guidelines for investigating causality of sequence variants in human disease, *Nature* 508:469–476, 2014.

Mademont-Soler I, Morales C, Soler A, Martinez-Crespo JM, Shen Y, Margarit E, et al: Prenatal diagnosis of chromosomal abnormalities in fetuses with abnormal cardiac ultrasound findings: evaluation of chromosomal microarray-based analysis, *Ultrasound Obstet Gynecol* 41:375–382, 2013.

Matthijs G, Souche E, Alders M, Corveleyn A, Eck S, Feenstra I, et al: Guidelines for diagnostic next-generation sequencing, *Eur J Hum Genet* 24:2–5, 2016.

McGillivray G, Rosenfeld JA, Gardner RJMK, Gillam LH: Genetic counselling and ethical issues with chromosome microarray analysis in prenatal testing, *Prenat Diagn* 32:389–395, 2012.

Meeuwesen L, Brink-Muinen A vd, Hofstede G: Can dimensions of national culture predict cross-national differences in medical communication? *Patient Educ Couns* 75:58–66, 2009.

Meir K, Fellig Y, Meiner V, Korman SH, Shaa A, Nadjari M, et al: Severe infantile carnitine palmitoyl-transferase II deficiency in 19-week fetal sibs, *Pediatr Dev Pathol.* 12(6):481–486, 2009.

Mellis R, Chandler N, Chitty LS: Next-generation sequencing and the impact on prenatal diagnosis, *Expert Rev Mol Diagn* 18:689–699, 2018.

Miller DT, Adam MP, Aradhya S, Biesecker LG, Brothman AR, Carter NP, et al: Consensus statement: chromosomal microarray is a first-tier clinical diagnostic test for individuals with developmental disabilities or congenital anomalies, *Am J Hum Genet* 86:749–764, 2010.

Miny P, Wenzel F, Tercanli S, Filges I: Chromosomal microarrays in prenatal diagnosis: time for a change of policy? *Microarrays* 2:304–317, 2013.

Monaghan KG, Leach NT, Pekarek D, Prasad P, Rose NC, ACMG Professional Practice and Guidelines Committee: The use of fetal exome sequencing in prenatal diagnosis: a points to consider document of the American College of Medical Genetics and Genomics (ACMG), *Genet Med* 22:675–680, 2020.

Mone F, Eberhardt RY, Morris RK, Hurles ME, McMullan DJ, Maher ER, et al: COngenital heart disease and the Diagnostic yield with Exome sequencing (CODE Study): prospective cohort study and systematic review, *Ultrasound Obstet Gynecol* 57(1):43–51, 2021.

Muys J, Jacquemyn Y, Blaumeiser B, Bourlard L, Brison N, Bulk S, et al: Prenatally detected copy number variants in a national cohort: a postnatal follow-up study, *Prenat Diagn* 40(10):1272–1283, 2020.

Normand EA, Braxton A, Nassef S, Ward PA, Vetrini F, He W, et al: Clinical exome sequencing for fetuses with ultrasound abnormalities and a suspected Mendelian disorder, *Genome Med* 10:74, 2018.

Novelli A, Cavalli P, Bernardini L: The future of prenatal diagnosis: karyotype, microarray or both? Technical and ethical considerations, *Expert Rev Proteom* 10:131–134, 2013.

Ogilvie CM, Yaron Y, Beaudet AL: Current controversies in prenatal diagnosis 3: for prenatal diagnosis, should we offer less or more than metaphase karyotyping? *Prenat Diagn* 29:11–14, 2009.

Olsen L, Sparso T, Weinsheimer SM, Dos Santos MBQ, Mazin W, Rosengren A, et al: Prevalence of rearrangements in the 22q11.2 region and population-based risk of neuropsychiatric and developmental disorders in a Danish population: a case-cohort study, *Lancet Psychiatry* 5:573–580, 2018.

Oneda B, Rauch A: Microarrays in prenatal diagnosis, *Best Pract Res Clin Obstet Gynaecol* 42:53–63, 2017.

Pangalos C, Hagnefelt B, Lilakos K, Konialis C: First applications of a targeted exome sequencing approach in fetuses with ultrasound abnormalities reveals an important fraction of cases with associated gene defects, *PeerJ* 4:e1955, 2016.

Petrovski S, Aggarwal V, Giordano JL, Stosic M, Wou K, Bier L, et al: Whole-exome sequencing in the evaluation of fetal structural anomalies: a prospective cohort study, *Lancet* 393:758–767, 2019.

Richards S, Aziz N, Bale S, Bick D, Das S, Gastier-Foster J, et al: Standards and guidelines for the interpretation of sequence variants: a joint consensus recommendation of the American College of Medical Genetics and Genomics and the Association for Molecular Pathology, *Genet Med* 17:405–423, 2015.

Riedijk S, Diderich KEM, van der Steen SL, Govaerts LCP, Joosten M, Knapen MF, et al: The psychological challenges of replacing conventional karyotyping with genomic SNP array analysis in prenatal testing, *J Clin Med* 3:713–723, 2014.

Riggs ER, Church DM, Hanson K, Horner VL, Kaminsky EB, Kuhn RM, et al: Towards an evidence-based process for the clinical interpretation of copy number variation, *Clin Genet* 81:403–412, 2012.

Riggs ER, Andersen EF, Cherry AM, Kantarci S, Kearney H, Patel A, et al: Technical standards for the interpretation and reporting of constitutional copy-number variants: a joint consensus recommendation of the American College of Medical Genetics and Genomics (ACMG) and the Clinical Genome Resource (ClinGen), *Genet Med* 22:245–257, 2020.

Robson SC, Chitty LS, Morris S, Verhoef T, Ambler G, Wellesley DG, et al: Evaluation of array comparative genomic hybridisation in prenatal diagnosis of fetal anomalies: a multicentre cohort study with cost analysis and assessment of patient, health professional and commissioner preferences for array comparative genomic hybridisation, *Efficacy Mech Eval*:2017.

Rooryck C, Toutain J, Cailley D, Bouron J, Horovitz J, Lacombe D, et al: Prenatal diagnosis using array-CGH: a French experience, *Eur J Med Genet* 56:341–345, 2013.

Rose NC, Anjali JK, Dugoff L, Norton ME, American College of Obstetricians and Gynecologists: Screening for fetal chromosomal abnormalities: ACOG practice bulletin, Number 226, *Obstet Gynecol* 136:e48–e69, 2020.

Rosenfeld JA, Coe BP, Eichler EE, Cuckle H, Shaffer LG: Estimates of penetrance for recurrent pathogenic copy-number variations, *Genet Med* 15:478–481, 2013.

Schmid M, Stary S, Springer S, Bettelheim D, Husslein P, Streubel B: Prenatal microarray analysis as second-tier diagnostic test: single-center prospective study, *Ultrasound Obstet Gynecol* 41:267–273, 2013.

Scott F, Murphy K, Carey L, Greville W, Mansfield N, Barahona P, et al: Prenatal diagnosis using combined quantitative fluorescent polymerase chain reaction and array comparative genomic hybridization analysis as a first-line test: results from over 1000 consecutive cases, *Ultrasound Obstet Gynecol* 41:500–507, 2013.

Silva M, Leeuw N de, Mann K, Schuring-Blom H, Morgan S, Giardino D, et al: European guidelines for constitutional cytogenomic analysis, *Eur J Hum Genet* 27:1–16, 2019.

South ST, Lee C, Lamb AN, Higgins AW, Kearney HM: ACMG standards and guidelines for constitutional cytogenomic microarray analysis, including postnatal and prenatal applications: revision 2013, *Genet Med* 15:901–909, 2013.

Srebniak M, Boter M, Oudesluijs G, Joosten M, Govaerts L, Van Opstal D, et al: Application of SNP array for rapid prenatal diagnosis: implementation, genetic counselling and diagnostic flow, *Eur J Hum Genet* 19:1230–1237, 2011.

Srebniak MI, Boter M, Oudesluijs GO, Cohen-Overbeek T, Govaerts LC, Diderich KE, et al: Genomic SNP array as a gold standard for prenatal diagnosis of foetal ultrasound abnormalities, *Mol Cytogenet* 5:14, 2012.

Srebniak MI, Mout L, Van D, et al: 0.5Mb array as a first-line prenatal cytogenetic test in cases without ultrasound abnormalities and its implementation in clinical practice, *Hum Mutat* 34:1298–1303, 2013.

Srebniak MI, Diderich KE, Govaerts LC, Joosten M, Riedijk S, Galjaard RJ, et al: Types of array findings detectable in cytogenetic diagnosis: a proposal for a generic classification, *Eur J Hum Genet* 22:856–858, 2014.

Srebniak MI, Van Opstal D, Joosten M, Diderich KE, de Vries FA, Riedijk S, et al: Whole-genome array as a first-line cytogenetic test in prenatal diagnosis, *Ultrasound Obstet Gynecol* 45:363–372, 2015.

Srebniak MI, Diderich KE, Joosten M, Govaerts LC, Knijnenburg J, de Vries FA, et al: Prenatal SNP array testing in 1000 fetuses with ultrasound anomalies: causative, unexpected and susceptibility CNVs, *Eur J Hum Genet* 24:645–651, 2016a.

Srebniak MI, Govaerts LC, Diderich KE, Joosten M, de Vries FA, Galjaard RJ, et al: Is prenatal cytogenetic diagnosis with genomic array indicated in pregnancies at risk for a molecular or metabolic disorder? *Genet Med* 18:307–308, 2016b.

Srebniak MI, Maarten FCMK, Polak M, Joosten M, Diderich KEM, Govaerts LCP, et al: The influence of SNP-based chromosomal microarray and NIPT on the diagnostic yield in 10,000 fetuses with and without fetal ultrasound anomalies, *Hum Mutat* 38:880–888, 2017.

Srebniak MI, Joosten M, Knapen M, Arends LR, Polak M, van Veen S, et al: Frequency of submicroscopic chromosomal aberrations in pregnancies without increased risk for structural chromosomal aberrations: systematic review and meta-analysis, *Ultrasound Obstet Gynecol* 51:445–452, 2018.

Stark Z, Gillam L, Walker SP, McGillivray G: Ethical controversies in prenatal microarray, *Curr Opin Obstet Gynecol* 25:133–137, 2013.

Stefansson H, Meyer-Lindenberg A, Steinberg S, Magnusdottir B, Morgen K, Arnarsdottir S, et al: CNVs conferring risk of autism or schizophrenia affect cognition in controls, *Nature* 505:361–366, 2014.

Suela J, López-Expósito I, Querejeta ME, Martorell R, Cuatrecasas E, Armengol L, et al: Recommendations for the use of microarrays in prenatal diagnosis, *Med Clin (Barc)* 148:328.e1–e28.e8.

Tang J, Zhou C, Shi H, Mo Y, Tan W, Sun T, et al: Prenatal diagnosis of skeletal dysplasias using whole exome sequencing in China, *Clin Chim Acta* 507:187–193, 2020.

Tetreault M, Bareke E, Nadaf J, Alirezaie N, Majewski J: Whole-exome sequencing as a diagnostic tool: current challenges and future opportunities, *Expert Rev Mol Diagn* 15:749–760, 2015.

Townsend A, Rousseau F, Friedman J, Adam S, Lohn Z, Birch P: Autonomy and the patient's right 'not to know' in clinical whole-genomic sequencing, *Eur J Hum Genet* 22:6, 2014.

van der Steen SL, Diderich KE, Riedijk SR, Verhagen-Visser J, Govaerts LC, Joosten M, et al: Pregnant couples at increased risk for common aneuploidies choose maximal information from invasive genetic testing, *Clin Genet* 88:25–31, 2015.

van der Steen SL, Riedijk SR, Verhagen-Visser J, Govaerts LC, Srebniak MI, Van Opstal D, et al: The psychological impact of prenatal diagnosis and disclosure of susceptibility loci: first impressions of parents' experiences, *J Genet Couns* 25:1227–1234, 2016.

van der Steen SL, Bunnik EM, Polak MG, Diderich KEM, Verhagen-Visser J, Govaerts LCP, et al: Choosing between higher and lower resolution microarrays: do pregnant women have sufficient knowledge to make informed choices consistent with their attitude? *J Genet Couns* 27:85–94, 2018.

Van El CG, Cornel MC, Borry P, Hastings RJ, Fellmann F, Hodgson SV, et al: Whole-genome sequencing in health care, *Eur J Hum Genet* 21:580–584, 2013.

van Nisselrooij AEL, Lugthart MA, Clur SA, Linskens IH, Pajkrt E, Rammeloo LA, et al: The prevalence of genetic diagnoses in fetuses with severe congenital heart defects, *Genet Med* 22:1206–1214, 2020.

Van Opstal D, de Vries F, Govaerts L, Boter M, Lont D, van Veen S, et al: Benefits and burdens of using a SNP array in pregnancies at increased risk for the common aneuploidies, *Hum Mutat* 36:319–326, 2015.

van Zwieten MC, Willems DL, Litjens LL, Schuring-Blom HG, Leschot N: How unexpected are unexpected findings in prenatal cytogenetic diagnosis? A literature review, *Eur J Obstet Gynecol Reprod Biol* 120:15–21, 2005.

Vanakker O, Vilain C, Janssens K, Van der Aa N, Smits G, Bandelier C, et al: Implementation of genomic arrays in prenatal diagnosis: the Belgian approach to meet the challenges, *Eur J Med Genet* 57:151–156, 2014.

Vears DF, Senecal K, Clarke AJ, Jackson L, Laberge AM, Lovrecic L, et al: Points to consider for laboratories reporting results from diagnostic genomic sequencing, *Eur J Hum Genet* 26:36–43, 2018.

Veltman JA, Brunner HG: Understanding variable expressivity in microdeletion syndromes, *Nat Genet* 42:192–193, 2010.

Vermeesch JR, Balikova I, Schrander-Stumpel C, Fryns JP, Devriendt K: The causality of de novo copy number variants is overestimated, *Eur J Hum Genet* 19:1112–1113, 2011.

Vermeesch JR, Paul DB, Sanlaville D, et al: Genome-wide arrays: quality criteria and platforms to be used in routine diagnostics, *Hum Mutat* 33:906–915, 2012.

Vetro A, Bouman K, Hastings R, McMullan DJ, Vermeesch JR, Miller K, et al: The introduction of arrays in prenatal diagnosis: a special challenge, *Hum Mutat* 33:923–929, 2012.

Vora NL, Gilmore K, Brandt A, Gustafson C, Strande N, Ramkissoon L, et al: An approach to integrating exome sequencing for fetal structural anomalies into clinical practice, *Genet Med* 22:954–961, 2020.

Wallis Y, Payne S, McAnulty C, Bodmer D, Sistermans E, Robertson K, et al: *Practice guidelines for the evaluation of pathogenicity and the reporting of sequence variants in clinical molecular genetics*, 2013, Association for Clinical Genetic Science (ACGS), Dutch Society of Clinical Laboratory Specialists.

Wapner RJ, Martin CL, Levy B, Ballif BC, Eng CM, Zachary JM, et al: Chromosomal microarray vs karyotyping for prenatal diagnosis, *N Engl J Med* 367:2175–2184, 2012.

Wapner RJ, Driscoll DA, Simpson JL: Integration of microarray technology into prenatal diagnosis: counselling issues generated during the NICHD clinical trial, *Prenat Diagn* 32:396–400, 2012.

Weiss MM, Van der Zwaag B, Jongbloed JDH, Vogel MJ, Brüggenwirth HT, Lekanne Deprez RH, et al: Best practice guidelines for the use of next-generation sequencing applications in genome diagnostics: a national collaborative study of dutch genome diagnostic laboratories, *Hum Mutat* 34:1313–1321, 2013.

Xue S, Yan H, Chen J, Li N, Wang J, Liu Y, et al: Genetic examination for fetuses with increased fetal nuchal translucency by genomic technology, *Cytogenet Genome Res* 160:57–62, 2020.

Yang K, Shen M, Yan Y, Tan Y, Zhang J, Wu J, et al: Genetic analysis in fetal skeletal dysplasias by trio whole-exome sequencing, *Biomed Res Int* 2019:2492590, 2019.

Yang X, Huang LY, Pan M, Xu LL, Zhen L, Han J, et al: Exome sequencing improves genetic diagnosis of fetal increased nuchal translucency, *Prenat Diagn* 40(11):1426–1431, 2020.

Zhou X, Wang Y, Shao B, Wang C, Hu P, Qiao F, et al: Molecular diagnostic in fetuses with isolated congenital anomalies of the kidney and urinary tract by whole-exome sequencing, *J Clin Lab Anal*: e23480, 2020.

CHAPTER 4

Dealing with uncertainty in prenatal genomics

Celine Lewis[1], Sam Riedijk[2] and Kelly E. Ormond[3]
[1]Population, Policy and Practice, UCL Institute of Child Health, London, United Kingdom
[2]Department of Clinical Genetics, Erasmus Medical Centre, Rotterdam, The Netherlands
[3]Department of Genetics and Stanford Center for Biomedical Ethics, Stanford School of Medicine, Stanford, CA, United States

What is uncertain in prenatal genomics?

Most prospective parents enter into the screening and diagnostic process with the assumption that the information they receive will be clear-cut and straightforward. In the face of potential risks, for example, advanced maternal age, a family history of a specific condition, or an abnormality identified on ultrasound, there are beneficent reasons to offer expanded testing to patients. Such testing may increase the diagnostic yield and minimize the diagnostic odyssey that patients would face postnatally. It can allow parents to plan for medical care, including in utero treatment, delivery planning, and perinatal palliative care as needed (Stark et al., 2013), and also can allow parents to make informed choices to continue or terminate the pregnancy when such options are legally available. And yet, testing can also pose risks, even those beyond the procedure itself. In addition, it can increase anxiety, provide information without clinical utility, and could lead to the termination of potentially healthy fetuses if there is a high degree of uncertainty in the results.

Some test results are relatively straightforward: for example, a karyotype obtained through amniocentesis or chorionic villus sampling that shows trisomy 21 has high analytic validity and clinical validity. Even then, there are prognostic ranges and uncertainties about what the lived experience will be for any specific child. There are many other types of results that can lead to uncertainty for parents. As testing has expanded to include broader tests [e.g., chromosomal microarray (CMA), panel testing, and exomes], the range of severity, penetrance, and clinical features that can be identified through testing has expanded to include conditions for which we have moderate prognostic clarity and others that are unknown. For example, a microduplication of 15q13.3 can be incompletely penetrant with variable expression, ranging from mild to severe neuropsychiatric symptoms that may include intellectual disability, behavioral impacts (e.g., bipolar disorder, Obsessive-compulsive disorder (OCD), schizophrenia), and epilepsy (Gillentine and Schaaf, 2015). Results can be broader than the specific

Prenatal Genetic Counseling
DOI: https://doi.org/10.1016/B978-0-12-823329-0.00016-7

indication for which the test is performed, including unexpected diagnoses related to the primary indication, or even incidental or secondary findings, such as the identification of a variant that increases the chance of cancer in adulthood. The more —genes that are examined, the higher the chance of obtaining a variant of unknown clinical significance (VUS) or identifying a variant in a gene of unknown significance. And finally, some percentage of potential parents are left with no explanation for the abnormal findings that prompted the testing in the first place.

Medicine is full of uncertainties, which arise from many different sources, and genetic testing is no different. One major source of uncertainty includes *probability*: of penetrance, of the frequency of certain features, and of the chances of survival. In genetics the limited knowledge base is often exacerbated by the small number of individuals affected by rare conditions. There is also an ascertainment bias, such that the most severely impacted individuals are most likely to be diagnosed, which potentially overestimates penetrance and expression of symptoms. In addition, fetal genotype–phenotype correlations are rarely available, and genes associated with more moderate risks have been recently identified, meaning that knowledge about the phenotype is newer and less well understood. Some conditions such as intellectual disability may not present prenatally, and the presentation of other conditions such as tuberous sclerosis may vary even within the same family. Other sources of uncertainty include limited knowledge about the condition because it is so rare, its natural history and prognosis, and the age of onset. Patients may be frustrated that their providers do not have much more information than is available to them by googling and on social media (Barton, et al., 2019). Additionally, diagnostic sensitivity may depend on the expertise of the health-care professional (HCP) (in the case of fetal imaging) and on the technology itself. Test limitations also present uncertainty such as the depth of sequencing, insufficient DNA material, or maternal cell contamination in the sample. And finally, there can be uncertainty about treatment efficacy.

These biases are especially impactful and challenging when considering the prenatal diagnosis of conditions where the natural history is limited by postnatal diagnosis, potentially exaggerating risk estimates. One example is the prenatal diagnosis of fetuses with 46,XX/46,XY (Amor, et al., 1999). This condition has been reported in childhood diagnoses of diseases of ambiguous genitalia, but those followed prospectively after prenatal diagnosis are most often phenotypically normal males.

Parents' experiences of uncertainty in the prenatal setting

Prospective parents usually enter into prenatal testing hoping to receive certainty and reassurance about their pregnancy. Generally, patients desire knowledge, imagining that "knowledge is power" and that genetic information would have clinically useful impact (Walser et al., 2015; Werner-Lin et al., 2016). However, they may be faced with information that is uncertain and difficult to navigate. A recent systematic review of the literature highlighted a myriad of uncertainties that parents had experienced (Harding et al., 2020). There were different sources of uncertainty, including those related to the procedure, diagnosis, and prognosis. There was the clinical impact of the uncertainty, including how parents struggled to make clinical decisions with the information they had been given, and finally, there was the emotional impact of the uncertainty, which included shock, worry, and decisional regret.

Some parents are unaware, even at the time of consenting to undergo testing, that they may receive inconclusive results, including VUS. In a qualitative study with couples who received uncertain CMA results, couples reported being "blind-sided" and unprepared to receive results for which clear risk estimates or diagnostic meaning were unknown (Werner-Lin et al., 2016). Where a result was given, parents experienced difficulty recalling the diagnosis (Walser et al., 2016) or there was very limited or imprecise information available about what the diagnosis meant prognostically, for example, whether it will result in the child having learning difficulties and, to what extent (Bernhardt et al., 2013; Rubel et al., 2017), or what the child will look like (van der Steen et al., 2015). In some cases, parents have been left uncertain around whether their baby will survive, creating a major source of anxiety (Hillman et al., 2013).

Receiving an uncertain result can significantly impact parents' emotional well-being. Parents have described experiencing shock, confusion, and worry at the time of receiving results (Werner-Lin et al., 2016; Walser et al., 2016; Bernhardt et al., 2013; Rubel et al., 2017; van der Steen et al., 2016; Werner-Lin et al., 2017; Quinlan-Jones et al., 2017; Wou et al., 2018). Bernhardt et al. (2013) described how in their interview study with women who had received abnormal CMA results, the women felt overwhelmed when receiving uncertain information, and discussed the difficulty of emotionally and intellectually managing the uncertainty of the results given how far into their pregnancy they were.

> *I think what is so difficult about these decisions—on top of the shock of it, the helplessness—is the timing...We felt like we were desperately trying to build enough information to make an informed decision in a very fast amount of time and that was very, very stressful for us.*

Some parents report regret about their decision to receive uncertain results (Bernhardt et al., 2013; Desai et al., 2018). Sometimes this type of knowledge is referred to as "toxic knowledge" because of the negative impact, it can have on the recipient (Bernhardt et al., 2013; Werner-Lin et al., 2019). Notably, a study conducted in Australia found that parents who chose to receive VUS results had higher decisional regret than those who chose not to receive these types of results (Halliday et al., 2018). Other longer term responses include loneliness, lingering anxiety and grief, feeling overwhelmed by the future, a lack of control over the situation, and feeling self-blame that have also been reported by parents (Werner-Lin et al., 2016;

Walser et al., 2016; Bernhardt et al., 2013; Rubel et al., 2017; Hillman et al., 2013; van der Steen et al., 2016; Werner-Lin et al., 2017; Quinlan-Jones et al., 2017; Wou et al., 2018; Halliday et al., 2018). In some cases, parents even minimize or avoid activities that might facilitate bonding with the baby (e.g., buying baby clothes or preparing the baby's nursery) because of the uncertainty around whether the baby will survive to term (Harding et al., 2020).

For those parents who continue with the pregnancy, anxiety may persist about the child's development, even after birth. In the study by Hammond et al. (2021), one parent, whose son was found to have a mass on his stomach during an ultrasound scan that later disappeared, commented that

> *for the first two years of his life, whenever he had tummy problems I thought 'oh, maybe it's that mass again.*

Similar findings have been identified elsewhere (Bernhardt et al., 2013; Werner-Lin et al., 2017; Desai et al., 2018; Govaerts et al., 2017), and this persistent anxiety mirrors that initially documented after false-positive newborn screening results (Sorenson, et al., 1984). A quantitative longitudinal study looking at the impact of receiving VUS on parental perception of child development found that parents who received VUS results rated their child as having more social-emotional and/or emotional problems and delays at 12 months than those parents who had received normal or likely benign results, although these differences were no longer statistically significant at 36 months (Desai et al., 2018).

Parents manage uncertainty in different ways. For some, it can be helpful to focus on what is known, and there is relief that certain known disorders have been ruled out. Others focus more on the unknown and the uncertainties generated by the result, for example, the risk of learning disabilities or structural malformations (Lou et al., 2020). Emotionally focused coping mechanisms such as remaining hopeful (Bernhardt et al., 2013; Werner-Lin et al., 2017) and making a conscious decision to suppress worry and focus on enjoying the pregnancy (Lou et al., 2020) can help parents. For some couples, it is important to see the child in its own right and not interpret the child's future development in light of the copy number variant (Lou et al., 2020). Action-focused coping strategies to manage uncertainty include pursuing increased medical and developmental assessments as a way to address long-term worry (Werner-Lin et al., 2017; Desai et al., 2018; Govaerts et al., 2017). Some access support groups or chat forums to speak with other parents who have had similar experiences (Wou et al., 2018; Hammond et al., 2021) and many search for information online or through additional clinical support (e.g., extra ultrasounds), although this sometimes creates additional anxiety or may result in misinformation (Hammond et al., 2021; Lou et al., 2020).

Health-care professionals' experiences of dealing with uncertainty in the prenatal setting

Genetic HCPs frequently facilitate informed consent and promote parent-centered decision-making in the clinic. But how do they experience pre- and posttest counseling around genomic sequencing-related uncertainties in the prenatal setting? A number of studies exploring the attitudes, experiences, and preferences of genetic HCPs have been published in recent years. A key finding from this work is that there are significant differences in attitudes and approaches to dealing with uncertain findings in the prenatal setting.

First, attitudes vary regarding *which* types of uncertain results HCPs believe should be communicated to parents. Some HCPs feel that where there is uncertainty about the future or actual significance and meaning of a result, it would be better not to share this information, while others are reluctant to withhold uncertain results from parents, even if it might have a negative impact on the doctor—parent relationship (Quinlan-Jones et al., 2016). Notably, in a study conducted in the Netherlands to determine expert consensus among prenatal experts on which chromosomal anomalies should and should not be returned to patients, there was disagreement on whether results with uncertain outcomes should be returned (Boormans et al., 2010). Attitudes are similarly mixed when looking at return of additional (e.g., secondary and incidental) findings. In a survey study conducted with genetic HCPs in the United States to determine their attitudes toward prenatal exome sequencing, the large majority of participants agreed that childhood-onset conditions with (93%) and without (80%) treatment should be returned to parents. However, attitudes regarding the return of results of adult-onset conditions were mixed, with only 50% of participants agreeing that conditions with available treatment should be returned (Brew et al., 2019).

Second, approaches vary regarding laboratory reporting practices for uncertain results. In a recent international cross-sectional study with HCPs dealing with uncertain CMA and/or exome sequencing results, there was variation in reporting practices both between and across countries for VUS, with some participants returning only pathogenic variants, and others returning some VUS. There was also variation in who decides what results are reported; in some laboratories, it is the clinical scientist, and in others, it is the clinician. However, there was broad agreement on reporting practices for incidental findings whereby only actionable or clinically significant results are returned (Lewis et al., 2021). Similarly, Vears, et al. (2017) conducted an international comparison study looking at whether or not laboratories reported VUS to clinicians. They found that some laboratories do not routinely report VUS to referring clinicians, but some will report VUS if they are thought to be relevant to the testing indication, but lacking sufficient evidence to classify them as pathogenic. Only a few laboratories would report VUS in candidate genes when they are not certain that they are relevant to the phenotype.

These findings highlight the complexity and contextual variation that exists, not only across countries with different health-care systems but also between HCPs themselves. The disparity in views likely reflects the tension between two competing professional duties: facilitating reproductive autonomy and avoiding unnecessary harm. They may also reflect a lack of consensus from professional bodies around the reporting of uncertain results, differences in national health systems (single-payer systems vs. mixed public/private providers), and/or cultural differences in how much information parents want and tolerance for uncertainty. Further research here would be valuable.

The question then turns to *why* there is reluctance among genetic HCPs to return uncertain results? HCPs have concerns when it comes to interpreting uncertain results in a prenatal setting, particularly where there is a lack of published data to support variant classification, or an extremely wide spectrum of clinical phenotypes (Durham et al., 2019; Bernhardt et al., 2014). This can lead to frustration for both the patient and the HCP who is generally regarded as an expert and accustomed to providing answers to parents' questions. HCPs are concerned that they may create unnecessary patient anxiety, as has been identified in a number of studies reporting parents' experiences of receiving uncertain results (Werner-Lin et al., 2016; Walser et al., 2016; Bernhardt et al., 2013; van der Steen et al., 2016; Werner-Lin et al., 2017; Quinlan-Jones et al., 2017; Wou et al., 2018). Horn et al. found that where there was uncertainty about the future or actual significance and meaning of results, HCPs and researchers were concerned that the uncertainty "could create more problems than [they are] trying to solve" (Horn and Parker, 2018). Receiving uncertain results may also put parents in a difficult situation where they either continue the pregnancy with the worry that the VUS implies the baby has a genetic anomaly or otherwise terminate, unsure whether the fetus was affected (Shkedi-Rafid et al., 2016). And finally, HCPs have raised concerns that uncertain information may lead to the termination of healthy pregnancies, and that is it difficult to supporting parents making decisions about whether to continue their pregnancy (Bernhardt et al., 2014; Mikhaelian et al., 2013).

Notable discrepancies may exist between providers' and parents' expectations for genomic sequencing and its results. In a qualitative study, genetic counselors acknowledged the low diagnostic yield of exome sequencing and their expectation that tests will often come back "normal" or uninformative, whereas they described that patients frequently overestimate the test's ability to reveal a diagnosis and are eager to consent (Park et al., 2019). This prompted counselors to modify patients' perceptions to reduce false hope through setting more realistic expectations around the diagnostic yield of the test, often through the use of probabilities. As described by one genetic counselor,

> *I think the most challenging thing with exome sequencing is balancing the patient's expectation that they're going to find an answer vs. . .the 30% chance that we're actually going to find the answer.*

As a result of the numerous clinical and ethical challenges associated with returning uncertain results, HCPs vary in whether they feel comfortable providing pre- and posttest counseling in this area. In a US study, nearly all genetic HCPs believed that there was potential for harm with prenatal exome sequencing due to ambiguous information and due to parents' misunderstanding the information provided by the test (Brew, et al., 2019). Bernhardt et al. (2014) found that when there is an uncertain CMA result, only 59% of genetic counselors felt comfortable providing genetic counseling to parents. Factors such as having sufficient knowledge of CMA, being familiar with current guidelines as well as having spent more time in prenatal practice, and believing that CMA is useful have all been found to be significant (Durham et al., 2019; Mikhaelian et al., 2013).

Managing uncertainty—the role of counseling

So how can HCPs best help patients manage prenatal uncertainty? First, ample research has shown that the quality of the relationship between patients and HCPs creates a sense of safety and improves treatment adherence, decision-making, and health and well-being outcomes (Biesecker, 2020; Kessler, 1999). Empathy (Limbo and Kobler, 2010) and kindness (Hutti et al., 2016) are essential to support and reinforce parents; empathic professionalism consists of the integration of professional knowledge, clinical experience, and the awareness that ones' own lived experience and personhood will color all of one's hypotheses and interactions (O'Connell et al., 2016). The definition of genetic counseling furthermore underscores the importance of psychoeducation (Semaka and Austin, 2019). Psychoeducation in prenatal counseling, for instance, implies educating expectant parents about the practicalities of decision-making, the vocabulary of grief, and the psychological process of going through a prenatal diagnostic trajectory. The therapeutic effects of psychoeducation may be feelings of relief from understanding that what one is experiencing is normal and reinforced autonomy through affirmation and validation. Additionally, psychoeducation may provide a sense of direction and overview, leading to an increased sense of control for patients (Evans, 2006).

Parents should specifically have the opportunity to take uncertainty into account in their decision-making process about prenatal WES. Parents highly appreciate honesty and transparency (Hammond et al., 2021), and many prior studies recommend that HCPs acknowledge uncertainty and its implications throughout the counseling process (Park et al., 2019; Kiedrowski et al., 2016; Newson et al., 2016). There is broad consensus about the benefits of being proactively transparent when HCPs convey to patients what is known and unknown in terms of genomics sequencing (Monaghan et al., 2020; Hegde et al., 2015), including addressing limitations of the test and uncertainties of test outcomes during pretest counseling. Clinician disclosure of uncertainty in these ways has been shown to improve trust and is a valuable aspect of patient-centered communication

(Henry, 2006; Kennedy, 2017; Lian and Robson, 2017). This type of genetic counseling approach requires sufficient self-confidence to convey what you as an HCP know, and also where the boundaries of knowledge and uncertainty exist. Ideally, counseling should be used to make the patients feel that they are part of the multidisciplinary team that works together behind the scenes and helps each other to deal with uncertainty.

There is a significant body of literature discussing the theoretical underpinnings of uncertainty in medical decision-making and relevant communication strategies (Simpkin and Armstrong 2019; Han et al., 2011). However, there are few guidelines providing insights into *how* to address these limitations, disclaimers, and uncertainties during counseling, despite HCPs expressing a need for guidance in this area (Reiff et al., 2013; Godfrey and Clark, 2014). A few specific suggestions are provided later and in Fig. 4.1.

Supporting parents through decision-making

While it may feel comfortable for the HCP to describe test limitations and uncertainties, parents may not possess the expertise to attribute meaning to these uncertainties. Additionally, the cognitive distractions and anxiety that occur during the prenatal diagnostic process may make it more difficult for patients to focus and make decisions regarding the course of the pregnancy. Being transparent around these issues provides a starting point to an important conversation, but the conversation should not end there. Many parents look to clinicians to provide direction: is there a real chance that genomic sequencing will uncover something clinically useful about the nature of the anomalies? Or will genomic sequencing create even more uncertainty? Although in principle these types of counseling sessions are nondirective (Kessler, 1999; Evans, 2006), many patients crave some directiveness when faced with uncertain information (Evans, 2006). Having the HCP apply clinical meaning—the clinician conveying his or her expectations about what genomic sequencing may (or may not) yield—can be a strong psychological intervention, providing some firm ground in the midst of a pool of diagnostic quicksand.

Clarifying what is and is not known

It can be useful for HCPs to think about guiding parents through decision-making in a concrete and step-by-step manner. They may want to start by clarifying what is and is not known. For example, given the fetal anomalies that were detected, what are the potential diagnoses being considered? What possible diagnoses can be ruled out based on the available test results? What has become more or less likely based on new information? Which information remains lacking and requires more investigation? Sometimes parents may perceive inconclusive or negative results as disappointing

Pretest

- Be aware that there are multiple sources of uncertainty linked to prenatal testing and parents differ in their willingness to be told uncertain results.
- Parents should be told of the potential for uncertain outcomes. This includes:
  - Uncertainties related to the results, for example, VUS, no genetic result, uncertain prognosis following clear genetic diagnosis;
  - Uncertainties related to the limitations of testing process, for example, diagnostic yield, potential for false negatives;
  - Uncertainties related to our current knowledge and understanding, for example, genotype–phenotype correlation.
- To manage the uncertainties related to test procedure, parents should be told all steps of the testing process, including how long they will have to wait for results and how they will be contacted.
- Parents will vary in their tolerance for uncertain information, and this may change over the course of the pregnancy. Ideally, health-care professionals should discuss patients' preferences for being told uncertain information prior to testing.

Posttest following an uncertain result

- Uncertain findings can be challenging to explain and as such flexibility may be needed for how long the posttest counseling session takes to allow time for explanations, discussing implications, and addressing questions from parents.
- Some nongenetics clinicians may need input from genetics specialists to explain uncertain findings and their implications.
- If VUS is reported, it is important to discuss the possibility that the classification of the variant may change as new knowledge is gained. In addition, policies for reanalysis when there are VUS or no significant findings should be made clear to parents.
- Parents may value being told what is known and certain as a way of relieving anxiety and establishing normality in pregnancy. This could include explaining that many syndromes and severe disorders have been ruled out through testing and/or through taking extra care to point out what is structurally normal at any follow-up ultrasounds.
- Parents who have been told uncertain findings may value psychological support, for example, through a genetic counselor or other psychological support service.
- Understand that parents may want to ask further questions after the initial posttest counseling session as they digest the information. Ensure that they have appropriate contact details to access follow-up care.
- Where appropriate, signpost parents to support groups including those for parents without a diagnosis.
- Give parents suggestions for how to talk to family members about the uncertain findings, including suggestions as to how to explain what is uncertain about their result.

Figure 4.1 Recommendations for pre- and posttest counseling about uncertainty. *Taken from: Harding E, Hammond J, Chitty LS, et al.: Couples experiences of receiving uncertain results following prenatal microarray or exome sequencing: a mixed-methods systematic review,* Prenat Diagn *40:1028–1039, 2020.*

news. In these cases the clinician should explicitly clarify that many conditions have been ruled out in the face of negative genomic sequencing results.

If possible, parents should be given the most likely outcomes of genome sequencing and what those outcomes would mean in terms of decision-making. For instance, the HCP could say something such as, "If no cause is found, then I would estimate the chance of these anomalies constituting a syndrome as rather low." Or "If we find that this is indeed the cause, then this diagnosis is associated with quite significant learning difficulties and many of these children will need support throughout their lives." Most children with this condition are unable to attend mainstream school. Parents can then consider: "What does it mean to us that our child will require ongoing support throughout their lifetime?."

Encouraging parents to consider their personal values and moral standpoint

HCPs can actively help parents consider the implications of the possible scenarios, including what personal meaning they attribute for each hypothetical scenario, and how they might feel if confronted with this situation. This may be challenging for parents but is a useful exercise in helping them clarify their views. Most parents find themselves defining quality of life, what they would consider an acceptable quality of life for their child, themselves, and the family unit, and then they critically assess their own mental and practical resources to determine whether they can offer their child all they feel that it deserves. For example, some parents place great importance on the prospect of a child going to mainstream school and living an independent life. Therefore it is important to propose a scenario where this might not be the case, to give parents the opportunity to consider how they might feel in such a scenario. Similarly, some parents indicate that they would be very willing to care for their child for as long as they are able to, but worry about what will happen to their child when they are gone, and may not want to burden their other children with this duty to care for their sibling in the future. Helping parents to consider these potential scenarios is, therefore, important. Parents often state that it is hard to make a decision based on uncertain grounds, as things may turn out to be different from the scenarios they have been considering. However, even scenarios that prove to be incorrect allow parents to start deliberating concepts such as the quality of life, the potential impact of care, their resources, the impact on others (e.g., other children or family members), and practical arrangements.

Finally, there are aspects of timing that make prenatal decision-making challenging for parents. Waiting for additional, potentially clarifying test results before engaging in the decision-making process may mean that parents lose valuable time in making a decision that is in line with their particular values and moral position. Encouraging parents to start considering hypothetical scenarios at the earliest opportunity is, therefore, important. Identifying essential considerations allows parents to regain a psychological grip. Counselors can facilitate these thought experiments by asking "what if" questions, paraphrasing and summarizing to get at personal meaning and values. In this manner, parents can disentangle their thoughts, feelings, moral considerations, and images of the future and weigh the importance of each in the configuration of their decision. This process of deliberation elucidates certainties and uncertainties, resources, and vulnerabilities and generates a sustainable fundament for reaching informed and future-proof decisions about prenatal diagnosis and pregnancy management.

In summary, pregnant couples can be confronted with numerous sources of uncertainty. This chapter has explored the different types of uncertainty that parents and HCPs may encounter through prenatal genomics, their views and experiences of these uncertainties, and how uncertainty may be addressed during the counseling process. The patient journey in the "Parents' experiences of uncertainty in the prenatal setting" section will also be useful in adding to our understanding of a prospective parent's experience. Additionally, it is important to be sensitive to cultural differences and one's own conceptions of cultural differences in counseling couples in prenatal genetics, which is the topic of the next chapter.

References

Amor D, Delatycki MB, Susman M, et al: 46,XX/46,XY at amniocentesis in a fetus with true hermaphroditism, *J Med Genet* 36:866–869, 1999.

Barton KS, Wingerson A, Barzilay JR, et al: Before Facebook and before social media...we did not know anybody else that had this: parent perspectives on internet and social media use during the pediatric clinical genetic testing process, *J Commun Genet* 10:375–383, 2019.

Bernhardt BA, Kellom K, Barbarese A, et al: An exploration of genetic counselors' needs and experiences with prenatal chromosomal microarray testing, *J Genet Counseling* 23:938–947, 2014.

Bernhardt BA, Soucier D, Hanson K, et al: Women's experiences receiving abnormal prenatal chromosomal microarray testing results, *Genet Med* 15:139–145, 2013.

Biesecker B: Genetic counseling and the central tenets of practice, *Cold Spring Harbor Perspect Med* 10:2020.

Boormans EM, Birnie E, Knegt AC, et al: Aiming at multidisciplinary consensus: what should be detected in prenatal diagnosis? *Prenat Diagn* 30:1049–1056, 2010.

Brew CE, Castro BA, Pan V, et al: Genetics professionals' attitudes toward prenatal exome sequencing, *J Genet Counseling* 28:229–239, 2019.

Desai P, Haber H, Bulafka J, et al: Impacts of variants of uncertain significance on parental perceptions of children after prenatal chromosome microarray testing, *Prenat Diagn* 38:740–747, 2018.

Durham L, Papanna R, Stevens B, et al: The utilization of prenatal microarray: a survey of current genetic counseling practices and barriers, *Prenat Diagn* 39:351–360, 2019.

Evans C: *Genetic counselling: a psychological approach*, Cambridge, 2006, University Press.

Gillentine MA, Schaaf CP: The human clinical phenotypes of altered CHRNA7 copy number, *Biochem Pharmacol* 97:352–362, 2015.

Godfrey E, Clark P: Developing standards for chromosomal microarray testing counselling in paediatrics, *Acta Paediatr* 103:574–577, 2014. (Oslo, Norway: 1992).

Govaerts L, Srebniak M, Diderich K, et al: Prenatal diagnosis of susceptibility loci for neurodevelopmental disorders – genetic counseling and pregnancy outcome in 57 cases, *Prenat Diagn* 37:73–80, 2017.

Halliday JL, Muller C, Charles T, et al: Offering pregnant women different levels of genetic information from prenatal chromosome microarray: a prospective study, *Eur J Hum Genet* 26:485–494, 2018.

Hammond J, Klapwijk JE, Hill M, et al: Parental experiences of uncertainty following an abnormal fetal anomaly scan: insights using Han's taxonomy of uncertainty, *J Genet Couns* 30:198–210, 2021.
Han PK, Klein WM, Arora NK: Varieties of uncertainty in health care: a conceptual taxonomy, *Med Dec Mak* 31:828–838, 2011.
Harding E, Hammond J, Chitty LS, et al: Couples experiences of receiving uncertain results following prenatal microarray or exome sequencing; a mixed-methods systematic review, *Prenat Diagn* 40:1028–1039, 2020.
Hegde M, Bale S, Bayrak-Toydemir P, et al: Reporting incidental findings in genomic scale clinical sequencing—a clinical laboratory perspective: a report of the Association for Molecular Pathology, *J Mol Diagn* 17:107–117, 2015.
Henry MS: Uncertainty, responsibility, and the evolution of the physician/patient relationship, *J Med Ethics* 32:321–323, 2006.
Hillman SC, Skelton J, Quinlan-Jones E, et al: "If it helps..." the use of microarray technology in prenatal testing: patient and partners reflections, *Am J Med Genet, A* 161A:1619–1627, 2013.
Horn R, Parker M: Health professionals' and researchers' perspectives on prenatal whole genome and exome sequencing: "We can't shut the door now, the genie's out, we need to refine it", *PLoS One* 13:e0204158, 2018.
Hutti MH, Polivka B, White S, et al: Experiences of nurses who care for women after fetal loss, *J Obstet Gynecol Neonatal Nurs* 45:17–27, 2016.
Kennedy AG: Managing uncertainty in diagnostic practice, *J Eval Clin Prac* 23:959–963, 2017.
Kessler S: Psychological aspects of genetic counseling: XIII. Empathy and decency, *J Genet Counseling* 8:333–343, 1999.
Kiedrowski LA, Owens KM, Yashar BM, et al: Parents' perspectives on variants of uncertain significance from chromosome microarray analysis, *J Genet Counseling* 25:101–111, 2016.
Lewis C, Hammond J, Klapwijk JE, et al. Dealing with uncertain results from chromosomal microarray and exome sequencing in the prenatal setting: An international cross-sectional study with healthcare professionals. Prenatal diagnosis. 2021. Mar 16. Available from: https://doi.org/10.1002/pd.5932.
Lian OS, Robson C: "It's incredible how much I've had to fight." Negotiating medical uncertainty in clinical encounters, *Int J Qual Stud Health Well-being* 12:1392219, 2017.
Limbo R, Kobler K: The tie that binds: relationships in perinatal bereavement, *MCN Am J Matern Child Nurs* 35:316–321, 2010. quiz 21–3.
Lou S, Lomborg K, Lewis C, et al: "It's probably nothing, but..." Couples' experiences of pregnancy following an uncertain prenatal genetic result, *Acta Obstet Gynecol Scand* 99:791–801, 2020.
Mikhaelian M, Veach PM, MacFarlane I, et al: Prenatal chromosomal microarray analysis: a survey of prenatal genetic counselors' experiences and attitudes, *Prenat Diagn* 33:371–377, 2013.
Monaghan KG, Leach NT, Pekarek D, et al: The use of fetal exome sequencing in prenatal diagnosis: a points to consider document of the American College of Medical Genetics and Genomics (ACMG), *Genet Med* 22:675–680, 2020.
Newson AJ, Leonard SJ, Hall A, et al: Known unknowns: building an ethics of uncertainty into genomic medicine, *BMC Med Genomics* 9:57, 2016.
O'Connell O, Meaney S, O'Donoghue K: Caring for parents at the time of stillbirth: how can we do better, *Women Birth* 29:345–349, 2016.
Park J, Zayhowski K, Newson AJ, et al: Genetic counselors' perceptions of uncertainty in pretest counseling for genomic sequencing: a qualitative study, *J Genet Counseling* 28:292–303, 2019.
Quinlan-Jones E, Hillman SC, Kilby MD, et al: Parental experiences of prenatal whole exome sequencing (WES) in cases of ultrasound diagnosed fetal structural anomaly, *Prenat Diagn*:2017.
Quinlan-Jones E, Kilby MD, Greenfield S, et al: Prenatal whole exome sequencing: the views of clinicians, scientists, genetic counsellors and patient representatives, *Prenat Diagn* 36:935–941, 2016.
Reiff M, Bernhardt BA, Mulchandani S: Genomic variation: what does it mean? *LDI Issue Brief* 18:1–4, 2013.
Rubel MA, Werner-Lin A, Barg FK, et al: Expert knowledge influences decision-making for couples receiving positive prenatal chromosomal microarray testing results, *Cult Med Psychiatry* 41:382–406, 2017.
Semaka A, Austin J: Patient perspectives on the process and outcomes of psychiatric genetic counseling: an "Empowering Encounter.", *J Genet Counseling* 28:856–868, 2019.

Shkedi-Rafid S, Fenwick A, Dheensa S, et al: What results to disclose, when, and who decides? Healthcare professionals' views on prenatal chromosomal microarray analysis, *Prenat Diagn* 36:252–259, 2016.

Simpkin AL, Armstrong KA: Communicating uncertainty: a narrative review and framework for future research, *J Gen Internal Med* 34:2586–2591, 2019.

Sorenson JR, Levy HL, Mangione TW, et al: Parental response to repeat testing of infants with 'false-positive' results in a newborn screening program, *Pediatrics* 73:183–187, 1984.

Stark Z, Gillam L, Walker SP, et al: Ethical controversies in prenatal microarray, *Curr Opin Obstet Gynecol* 25:133–137, 2013.

van der Steen SL, Diderich KE, Riedijk SR, et al: Pregnant couples at increased risk for common aneuploidies choose maximal information from invasive genetic testing, *Clin Genet* 88:25–31, 2015.

van der Steen SL, Riedijk SR, Verhagen-Visser J, et al: The psychological impact of prenatal diagnosis and disclosure of susceptibility loci: first impressions of parents' experiences, *J Genet Counseling* 25:1227–1234, 2016.

Vears DF, Sénécal K, Borry P: Reporting practices for variants of uncertain significance from next generation sequencing technologies, *Eur J Med Genet* 60:553–558, 2017.

Walser SA, Kellom KS, Palmer SC, et al: Comparing genetic counselor's and patient's perceptions of needs in prenatal chromosomal microarray testing, *Prenat Diagn* 35:870–878, 2015.

Walser SA, Werner-Lin A, Russell A, et al: "Something Extra on Chromosome 5": parents' understanding of positive prenatal chromosomal microarray analysis (CMA) results, *J Genet Counseling* 25:1116–1126, 2016.

Werner-Lin A, Barg FK, Kellom KS, et al: Couple's narratives of communion and isolation following abnormal prenatal microarray testing results, *Qual Health Res* 26:1975–1987, 2016.

Werner-Lin A, McCoyd JLM, Bernhardt BA: Actions and uncertainty: how prenatally diagnosed variants of uncertain significance become actionable, *Hastings Cent Rep* 49(Suppl 1):s61–s71, 2019.

Werner-Lin A, Walser S, Barg FK, et al: "They Can't Find Anything Wrong with Him, Yet": mothers' experiences of parenting an infant with a prenatally diagnosed copy number variant (CNV), *Am J Med Genet, A* 173:444–451, 2017.

Wou K, Weitz T, McCormack C, et al: Parental perceptions of prenatal whole exome sequencing (PPPWES) study, *Prenat Diagn*:2018.

CHAPTER 5

Addressing diversity in prenatal genetic counseling

Ingrid A. Peters[1,2] and Fadua El Bouazzaoui[3]
[1]Obstetrics and Gynaecology, Erasmus University Medical Center, Rotterdam, The Netherlands
[2]Quality and Patient Care Erasmus University Medical Center, Rotterdam, The Netherlands
[3]Foundation Health Educators, Rotterdam, The Netherlands

Introduction

Because of their welfare and relatively economic and political stability, the postindustrial countries in the world are preferred by immigrants, which results in the increasing diversity of the population of these countries in terms of multiethnicity. It also resulted in a diverse composition of the reproductive women's population, which asks for the competence and knowledge about cultural, religious, and ethnic diversity, which is conditional for the provision of high-quality and value-based healthcare by reproductive caregivers, including genetic counselors.

In comparison with native pregnant women, immigrants are at a higher risk of adverse pregnancy outcomes and insufficient prenatal health-care utilization (Poeran et al., 2013; Pimentel and Eckardt, 2014). Also consanguinity is more often seen within migrant couples, which makes genetic counseling more needed (Posch et al., 2012).

Especially for immigrant pregnant women, a prenatal diagnosis of a fetal anomaly leads to an unexpected personal tragedy due to more compelling religious and cultural beliefs and convictions (Gitsels et al., 2014a,b; Carlsson et al., 2019). Tailored prenatal genetic counseling given in a respectful and empathically psychological way, addressing cultural and religious values, is desirable to meet the needs of pregnant women with an immigrant background (Carlsson et al., 2019). For prenatal genetic counselors, it is recommendable to understand the sociocultural and religious underpinnings of immigrant pregnant women regarding prenatal diagnostics as a first step to engage these women in shared and informed decision-making.

This chapter will inform about addressing diversity in prenatal genetic counseling, starting with an exploration of the concepts of cultural, religious, and ethnic differences. Further, an elaboration of the existing inequalities in prenatal healthcare and experiences with provided care within a multiethnic women's population is provided, followed by best practices for tailored prenatal genetic counseling with specific attention for *themes to discuss during counseling, communication, the role of the interpreter, health*

Prenatal Genetic Counseling
DOI: https://doi.org/10.1016/B978-0-12-823329-0.00006-4

education, and community-based *practical instructions*. The chapter will end with a self-assessment to test the gained knowledge.

Concepts of cultural, religious, and ethnic differences

The position, beliefs, and behavior of a pregnant woman and her social network are in varying degrees influenced by ethnicity, and cultural background and religious beliefs. Knowledge about these concepts and the influence on the individuality of the patient can provide the prenatal genetic counselor starting points for an open conversation about assumptions, beliefs, and attitudes toward genetic screening and diagnostics. Genetic counseling based on health education and decision-making support, which is, when needed, adjusted to the knowledge, and cultural and religious background of the pregnant women, contributes to the development of a trustful patient relation and a value-based care (Martin et al., 2013; Bastemeijer et al., 2017). Since concepts of culture, religion, and ethnicity have a related connection, it is first necessary to frame these background characteristics separately.

Ethnicity

Ethnicity frequently features in the political, healthcare, and research context, but it is rarely theorized (Zagefka, 2009). Popular theories might often suggest that ethnicity is grounded in biological differences between individuals and groups, but from the perspective of social psychology and epidemiology, ethnicity is defined as a social construct (Zagefka, 2009; Ford and Harawa, 2010).

Ethnicity is a context-specific, multilevel (group or individual), and multifactorial social construct that is related to race. The concept of ethnicity is used to differentiate diverse subgroups within a population of a specific region. The largest ethnic group within the population is called the majority and smaller ethnic groups have been called minority ones (Ford and Harawa, 2010).

As conceptualized by Ford and Harawa (2010), ethnicity can be defined as a two-dimensional, context-specific social construct (Ford and Harawa, 2010). First is an attributional dimension group characteristic, for example, culture and nativity and second is a relational dimension that indexes a group's position within a social hierarchy (minority vs majority ethnic group).

The societal context in which people live determines whether they can be categorized in a specific ethnic group. Individuals in the same ethnic group share a same set of sociocultural characteristics. Factors, for example, cultural and religious habits and behavior and being part of a numeric minority reinforce ethnicization of an individual or group (Zagefka, 2009; Ford and Harawa, 2010).

Culture

Williams (1983), one of the founders of cultural studies, underlines in his publications that the concept culture is one of the most complex-to-define "words" in English. The meaning of culture must be conceptualized from different perspectives (Williams, 1983; Atkinson, 2004). In this paragraph, we elaborate the concept of culture based on the theory of Hofstede (2010).

During the socialization of an individual (group), culture is transferred to the developing child and young adult (Hofstede, 2010). According to Geertz (1973), culture gives meaning to the world and makes it understandable from the perspective of an individual or group (Geertz, 1973). Notwithstanding the fact that cultural behavior is not always notable, all human beings are part of a specific cultural group. Culture is a phenomenon that is shared between individuals that live or lived in the same social environment (Hofstede, 2010). Culture can be defined as a specific collective mental programming of a group of individuals that differentiates that group when compared with other cultural ones (Hofstede, 2010). Based on the theory of Hofstede (2010), culture is built on five basic components: *values*, *norms*, *rituals*, *heroes* and *symbols*. *Values* are defined by the value or meaning that is connected with a specific material, event, or behavior. *Norms* consist of the behavioral rules within a cultural group. *Rituals* are specific cultural acts or proceedings that are prescriptive or especially forbidden within a cultural group. Within a culture that is fictive or real, individuals or entities can have a valuable position and are framed as *heroes* within a group culture. *Symbols* are signs and objects that have a large meaning within culture.

These components are expressed in cultural behavior and have a different priority and position within the expression and practice of a specific culture. As shown in Fig. 5.1 (religious), *values* function as the core influencing component of a culture, which determines cultural behavior concerning *rituals* and the definition of *heroes* within the culture. The use of *symbols* within a culture group, for example, wearing a scarf, is based on Hofstede's Cultural Onion, the most surface element of culture and usually the most recognizable for outsiders (not members of the cultural group) like health-care professionals.

Example of cultural behavior

To explore the concept of the culture of pregnant women within the setting of prenatal genetic counseling, an example of the five components of culture from the perspective of a North African (Moroccan) pregnant woman follows next.

The situation

Fadma, 29 years old, is pregnant with her second child. She has a first-generation Moroccan immigrant background, is married, and has an intermediate educational attainment level.

(Continued)

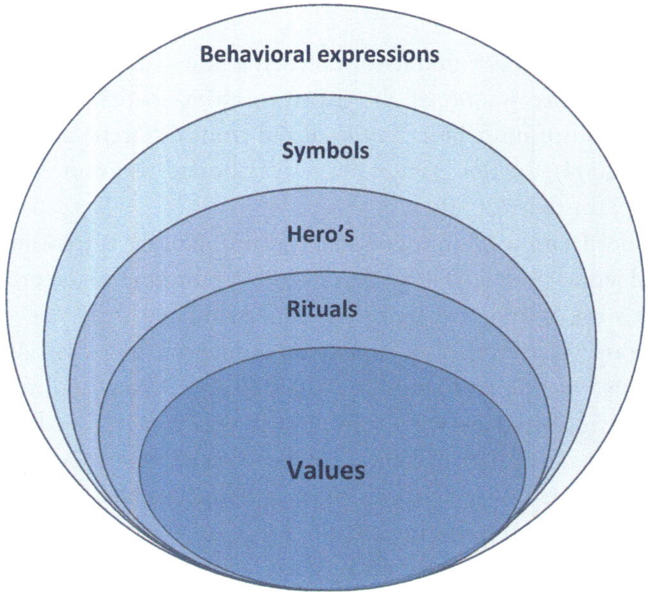

Figure 5.1 Hofstede's cultural onion.

(Continued)

She had a fetal anomaly scan at a gestational age of 20 weeks and 2 days. The scan shows structural anomalies that indicate that she carries a child with Edwards' syndrome. The sonographer and the midwife give a short posttest counseling and refer her for an advanced ultrasound scan in the hospital. The scan confirms the previous findings and the sonographer refers her to a clinical geneticist for prenatal genetic counseling. Decisions about participation in invasive prenatal diagnostics and continuation or termination of the pregnancy must be made.

Values

Fadma believes that her pregnancy is given from the hands of Allah. She considered not to participate in the first trimester prenatal screening for Down syndrome, Edwards' syndrome, and Patau's syndrome, because she accepts her unborn child with or without possible congenital disorders. She undergoes the second trimester fetal anomaly scan, because she thought it was a part of the standard care and she wants to know the gender of her unborn child.

The unfavorable outcome of the ultrasounds causes an inner struggle for Fadma. Before the test she did not take into account the unexpected outcome of the prenatal screening and diagnostic ultrasound scans. She doubts if she wants to go to the clinical geneticist for prenatal genetic counseling but finally decides to go. The clinical geneticist describes the

(Continued)

(Continued)

characteristics of the syndrome and the possibility to definitely diagnose (amniocenteses) the syndrome within this pregnancy. Fadma experiences the given information as harsh. She does not want to hear such worrying information about her unborn child. The possibility for participation in prenatal invasive diagnostics feels like an illogical option. Does she want more information or is it better to wait and see what the lessons of Allah will be? Based on her personal values, it feels better to do nothing at all. Her husband convinces her to undergo the amniocenteses, using a little pressure by telling a story about his nephew having multiple disabilities. This invasive test confirms a triploidy on chromosome 18, but she decides to continue her pregnancy. Despite the uncertainty of her husband about this choice, she is convinced about her decision based on religious values.

Norms

Within her family and social environment, it is not done to undergo prenatal screening or prenatal diagnostics. This is based on the assumption that these prenatal tests by definition lead to the termination of pregnancy when the unborn child is diagnosed with an anomaly.

Fadma made a not-informed decision about undergoing the fetal anomaly scan and after the outcome of the test, she must now consider whether or not to accept the referral for follow-up tests and prenatal genetic counseling. She doubts if it is appropriate within her culture to undergo diagnostic tests and if it is wise to get herself in a position that the doctor at some point asks her if she wants to end the pregnancy. After motivating conversations with her husband and sister in law, she decides to undergo the advanced ultrasound scan, go to the prenatal genetic counseling, and consult and participate in the amniocenteses. Especially the approval of her husband for the participation in the prenatal diagnostics was important for Fadma.

After the confirmation of Edwards' syndrome by the amniocenteses, a second consultation by the clinical geneticist follows. In this meeting, Fadma stated clearly that she wants to continue her pregnancy based on her religious convictions and behavioral rules. She recently consulted the Imam to discuss what to do when the outcome of prenatal diagnostics was unfavorable.

Heroes

Within Fadma's cultural group the Imam has an important position as a wise man who has the knowledge about the Qur'an and the Sunnah (the way in which prophet Mohammed lived his life, which is for Muslims within Fadma's cultural group an example of adequate behavior in specific situations of life). Fadma's choice for continuing the pregnancy was mainly based on the opinion of the Imam that termination of pregnancy is prohibited after 4 months when there is no life-threatening danger for the mother.

Rituals

From the moment Fadma knew she was pregnant, she always ate what she liked. She was influenced by the assumption within her cultural group that if she does not eat what

(*Continued*)

(Continued)

she likes when she is hungry, it can lead to the malformation of the unborn child. After the diagnosis that her child had Edwards' syndrome, she asked herself for a short moment if she lacked eating a specific kind of food.

Symbols

From the start of her pregnancy, Fadma wears a necklace with the eye of Fatima, because she believes that it protects her unborn child from "the evil eye." Although she knows that it is a cultural-based superstition and forbidden in the context of her Islamic religion, she finds the use of this symbol important, especially in the period of uncertainty during this pregnancy.

Collective versus individualistic culture

There are numerous cultures and subcultures. Even within a country, there are significant differences in culture when comparing urban areas with rural areas. To be able to make a distinction between cultures a differentiation can be made in culture dimensions as, for example, among others, the *collective* and *individualistic* cultures and *feminine* and *masculine* cultures (Hofstede, 2010). These cultural dimensions differ when comparing countries (Hofstede_Insights, 2020). Characteristics of these four categories of culture are as follows:

- *Collective culture* is a "we culture." The interest of the group holds the most importance within this culture. Group cohesion is strong within this culture and indirect communication with the avoidance of confrontation is common.
- *Individualistic culture* is an "individual culture." Members of this culture group care for themselves and only for their direct family. A direct communication style is common.
- *Feminine culture* is based on the assumption that having a high quality of life is a sign of a successful life. Equality and solidarity are important elements in this culture. Conflicts are resolved by compromises and negotiation.
- *Masculine culture* is characterized by the urge for competition, performance, dedication, and the pursuit to have success. Conflicts can be resolved by fighting them out.

Religion

Worldwide, there are numerous religions and spiritual traditions with their submovements. It is impossible to give a standard list with all religions and spiritual traditions and their appearances, because these are dynamic and constantly changing due to globalization and migration. Like with the concepts of culture and ethnicity, it is hard to define religion. Geertz (1973) provided an internationally accepted definition of religion in religion studies:

> *(1) a system of symbols which acts to (2) establish powerful, pervasive, and long-lasting moods and motivations in men by (3) formulating conceptions of a general order of existence and (4) clothing these conceptions with such an aura of factuality that (5) the moods and motivations seem uniquely realistic.*

The four basic functions of religion are as follows (Macionis and Plummer, 2011):

1. social cohesion: bonding between individuals within a group;
2. social control: behavioral norms;
3. to provide meaning: values for what is wrong or good;
4. purpose of life.

Abortion and religion

Islam

The rulings and regulations of Islam are based on the protection of basic human rights and the preservation of one's life, possessions, and intellect. These are fundamental principles and remain the same in different cultures and times.

The first two sources of the Islamic rulings and regulations are the Qur'an and the Sunnah. The Qur'an is deemed to be the literal word of God (Arabic: *Allah*). To understand the Qur'an the Muslims rely on the interpretation of the Qur'an by the Prophet Muhammad first, and then by his companions as they witnessed the revelation firsthand and knew what the verses were about. This explanation of the Qur'anic verses is collected in the exegesis of the Qur'an, which is called the *Tafsir*. There are various Tafsirs written over time, the most well known being the Tafsir of Ibn Kathir and the Tafsir of Tabari (Kathir, 2003).

The second source of the Islamic rulings is the Sunnah of the Prophet Muhammad. The Sunnah consists of what the Prophet said, what he did, and silently accepted. These are written down in the form of short stories narrated by the companions of the Prophet; these short stories are called *Hadith*. The most renowned Hadith collections are Sahih al-Bukhari and Sahih Muslim.

When it comes to the interpretation of the Qur'an and the Sunnah, and rulings of modern-day issues which were not present in the time of the Prophet Muhammad, Muslims refer to the Islamic scholars for their verdicts and explanation. This was commanded in the Qur'an: "So ask the people of the knowledge if you do not know" (The Noble Quran 21:7) and "Had they only referred it back to the Messenger and back to those in authority among them so that those of them who know how to extract it (the ruling) would have known it..." (The Noble Qur'an 4:83). Al-Tabari mentioned in his Tafsir that various prominent companions of the Prophet, like Ibn 'Abbas, said that "those in authority" mean the scholars. Ibn Kathir stated in his Tafsir that "those in authority" refer to the scholars and the Islamic rulers. These verdicts issued by Islamic scholars are called *fatawa* (singular: *fatwa*) and are only to be issued by senior Islamic scholars who are known for their broad knowledge and long study of various Islamic topics.

Generally, abortion is not permitted in Islam except due to an Islamically justifiable reason. Such reasons include endangerment of the health of the mother or pregnancy out of wedlock. In the second example the justification is the covering of a sin. Fear of poverty or wishing to limit the number of children that one has is not considered justifiable reasons.

Some Islamic scholars have stated that an abortion is only allowed up to 40 days after conception, while other scholars say that an abortion is allowed up to 17 weeks after conception. According to Islam, the soul is blown into the fetus after 17 weeks, and this is when the fetus is considered a person. For this reason the favored stance on this issue is the abortion may take place until 17 weeks after conception.

After 17 weeks, abortion is prohibited except if carrying through with the pregnancy is life threatening for the mother. In this case the life of the mother takes precedence over the life of the child and the pregnancy may be terminated (Khadeejah, 2020).

Other religions

Christians base their beliefs on the Gospels in the Bible. How the Bible is explained and implemented in daily life varies between different Christian groups. In Christianity the view on abortion varies between the different Christian denominations, and even within a denomination, opinions can differ, or views change over time. Within Christianity, views range from antiabortion to prochoice.

The Jewish Encyclopedia makes a distinction between the teachings of Moses and Rabbinic Judaism (The Jewish Encyclopedia, 1906). The second is considered mainstream Judaism in the present day. Rabbinic Judaism is based on the written law (Thora) and oral law (that was later written down in the Talmud). The written law alone was seemed insufficient as a guide for Jewish life, as the commandments needed further explanation of how to be carried out (Telushkin, 1991).

There is some difference within Judaism, as in all religions, about the legality of abortion. In general, abortion is not deemed permissible. In the case, carrying the fetus is life-threatening for the mother, an abortion is obligated. The abortion of a child out of wedlock is forbidden.

Examples of culture and religious assumptions and beliefs during pregnancy

There are a number of beliefs and assumptions surrounding the pregnancy in different cultures and religions. Each culture has its own assumptions. A lot of the assumptions that are deemed religious stem in actuality from culture. In the modern day and age, where science has made major advancements, we should expect for baseless assumptions and beliefs to vanish, but some beliefs are still present and can influence the thought process, emotions, and actions of pregnant women around the world.

One belief for instance is that a pregnant woman should eat whatever she craves; this belief is shared by various cultures, from Hispanic to Middle East (Alcaniz, 2003). In some Middle East countries, it is believed that a child takes his birth with a mark in the shape of the food his mother craved, if her craving is not fulfilled. As a result, women eat everything they feel like, which can lead to unhealthy eating habits. An example of a belief, which is persistent in Western countries today, is the belief that the shape of the pregnant woman's stomach can give away the sex of the unborn child.

Case 1

Maudy, 39 years old, Surinamese descent, born in the Netherlands

During my first pregnancy, I did not use prenatal screening. My son was born with PDD-NOS, which is an autism spectrum disorder. In my culture, we say when a couple argues during pregnancy, the chance the child is born with a disability increases. During my pregnancy, my partner and I separated. I have long regretted not using prenatal screening of my son and I felt guilty his disability could be my fault.

During my second pregnancy, I did use prenatal screening. During the screening the midwife explained everything looked fine. She told me prenatal screening tests could identify certain birth defects, but disorders like PDD-NOS could not be identified. She also said I was not to blame for my son's disorder. When I learned this, I felt relieved.

My experience taught me that every pregnant woman has the right and the duty to check information based on cultural myths.

Case 2

Salima, 32 years old, Moroccan descent

When I was 3-month pregnant, an ultrasound was made. It did not look well, as I had too much amniotic fluid in the womb. I was sent to the hospital for further examination. Upon examination the doctor determined that the baby had a chromosome abnormality. To determine which abnormality it was, further examination was necessary. After the amniotic fluid test, it turned out that the baby had Down syndrome. She also possibly had a heart condition. The prognosis was not good.

The gynaecologist said, "I could have the baby aborted up to 24 weeks of pregnancy." He said I could have complications during the pregnancy and birth if I did not have the fetus aborted, which would have an impact on my mental well-being. The doctor urged us to consider an abortion. We refused though. Due to our religious beliefs, we did not want to abort the baby. If the baby would not make it, this would be God's will. And the Qur'an states, "Allah burdens not a person beyond his scope" (Chapter 2, verse 186). Because the doctor did not support us in carrying through with the pregnancy, we requested to see a different doctor.

It was a difficult time. When I searched the internet, I only found discouraging information. I was disappointed by the lack of support I received at the hospital. I prayed to God to make things easy for me. My family and the Islamic community did support me in my decision not to have an abortion. I noticed within our community, people accepted that this was the Will of God.

(Continued)

(Continued)

After 6 months of pregnancy, it appeared that the extra fluid was gone. The heart of the baby also looked fine. After birth the baby only had Down syndrome, she did not have birth defects. I am grateful for my beautiful little girl. She was meant to be with me. When I look at my daughter now, it upsets me how the first gynecologist pressured us to have an abortion. Sometimes I want to go to the doctor and tell him how happy I am with the child he wanted me to abort.

I noticed people with the same cultural background as me sometimes do not know what to say when they hear a child has Down syndrome. Or when they have a child with Down syndrome, they are ashamed and cannot accept it. This can result in awkward situations. Once for instance, I saw another child with Down syndrome at the playground. I walked over to the mother and told her my child had Down's syndrome too. To my surprise the mother denied the child had Down syndrome. I think it is sad to be ashamed of your child and that is a sin.

I see my child as a blessing and a trial for me. I believe that God tests those He loves. I am happy to have her in my life. My religion helps me. My advice to medical professionals is to respect the choice of a mother, whatever her cultural or religious background may be.

Inequalities in prenatal care within a multiethnic women's population

Prenatal health-care provision

The individual characteristics of the pregnant woman influence the provision of obstetric care and prenatal genetic counseling. Especially, a low or a high age, a low educational level, a nonnative ethnicity, a low level of knowledge (health illiteracy), uninformed decision-making, and a negative attitude of the pregnant woman toward participation in prenatal test are negative determinants (Premkamur, 2009; Kalengayi et al., 2012; Boerleider et al., 2013a).

Several studies underline the fact that information provision about prenatal screening and diagnostics, consisting of counseling and providing leaflets only, is not sufficient for pregnant women with a non-Western immigrant and low social economic status background (Gal and Prigat, 2005; Wiegers et al., 2015; Carlsson et al., 2019). This leads more frequently to insufficiently informed decisions concerning participation in prenatal screening and diagnostics (Tschudin et al., 2011; Agt et al., 2012; Carlsson et al., 2019). While other research shows that pregnant women with an immigrant background have the desire to learn about prenatal diagnostics to make decisions about participation in prenatal screening and genetic testing (Garza et al., 2020).

In a study that explored the narratives from immigrants following an adverse prenatal diagnosis, three trends were observable: (1) the women experienced an unexpected hurricane of emotions, (2) they tried to understand the situation based on information in an unfamiliar language, and (3) being cared for in a country with accessible obstetric care and where induced abortion is legal (Carlsson et al., 2019). Previous research calls for improvements in the information provision procedure concerning prenatal screening and diagnostics when it comes to counseling and used information materials to bridge language proficiency, culture, religion, and health illiteracy barriers (Fransen et al., 2010; Gitsels et al., 2014, 2014a,b; Carlsson et al., 2019).

In addition, miscommunication in health-care communication to pregnant women with an immigrant background is more often present due to the fact that these women more often have to deal with an inadequate language proficiency level, a low educational level, health illiteracy, too short medical consults, comorbidity, differences in (healthcare) expectations, and limited trust between pregnant women and the care provider (Akhavan, 2012; Aquino et al., 2015). Understanding genetic tests, weighing diagnostic results, and making an informed choice about participating can be challenging for pregnant women and is especially difficult for those who have insufficient language proficiency as well as health illiteracy (Cheng et al., 2018; Jun et al., 2018). A study focusing on miscommunication in genetic counseling identified five common sources of miscommunication: (1) medical jargon; (2) nondirective nature of counseling; (3) misplaced cultural sensitivity; (4) problems of translation; and (5) problems of trust (Browner et al., 2003).

Prenatal health-care utilization

Inadequate use of prenatal care is more often seen within pregnant women with an immigrant background. Demographic, pregnancy, and socioeconomic factors as well play a role in this insufficient use of prenatal healthcare. Especially pregnant immigrant women with an age <19 and >36 years, living in a highly urban area, having an insufficient language proficiency, with an unexpected pregnancy, a low educational level, low income, and absence of a health insurance are at higher risk for inadequate use of care during pregnancy (Bennett et al., 2006; Boerleider et al., 2015; Posthumus et al., 2016; Heetkamp et al., 2020). The more common late start of prenatal care (≥11 weeks gestational age) of women with a non-Western, first-generation, immigrant background contributes to this underutilization of prenatal care (Alderliesten et al., 2007; Boerleider et al., 2015; Heetkamp et al., 2020). The before mentioned nonmedical determinants also play an influencing role in the delay of prenatal care (Alderliesten et al., 2007; Boerleider et al., 2015; Heetkamp et al., 2020). Also the responsiveness of the prenatal health-care provision procedure influences the utilization of care (Peters et al., 2019). Specific prenatal health-care organizational factors like location and setting, the role of various care providers (general practitioners, midwives, gynecologist, and other medical specialists), how appointments can be made, and the means of communication also have influence (Schouten and Meeuwesen, 2006; Kalengayi et al., 2012). Schouten and Meeuwesen (2006) assume that a present language barrier in the communication between the prenatal health-care professional and the patient does not subsequently lead to health-care utilization problems. Also other studies underline that the existing migration background of pregnant women has a more determining effect on the use of prenatal care (Boerleider et al., 2013a,b).

Also cultural factors may influence the use of prenatal diagnostics, as mentioned earlier in this chapter under "example of cultural behavior." Examples of these factors are convictions concerning health, lifestyle, and provided healthcare (use), related rituals and behavior, culturally determined social and family pressure, and dynamics (Kalengayi et al., 2012; Boerleider et al., 2013b; Gitsels et al., 2014). More specific, having a non-Western immigrant background is often associated with a lower uptake for prenatal screening and diagnostic tests and religious (Islamic/Protestant Christian) pregnant women, and women with a low language proficiency level have a lower uptake rate for the fetal anomaly scan (Agt et al., 2012; Gitsels et al., 2014a,b; Borges et al., 2016). Previous research shows that due to incomplete knowledge of religious doctrines, Islamic pregnant women were unaware that within Islamic tradition there is the possibility of termination if a fetus has serious anomalies (Gitsels et al., 2014a,b). This can influence both decisions of prenatal screening and diagnostic tests uptake and of terminating a pregnancy for fetuses with serious anomalies (Tschudin et al., 2011; Gitsels et al., 2014a,b; Garza et al., 2020).

Practical examples

The following cases are of three women from Islamic countries. The cases illustrate the differences between women, despite them sharing the same religious background. Religion is one part of various elements that can play a part in the woman's decision to choose for prenatal screening and consider terminating a pregnancy. It is important when treating a patient from an immigrant background not only to keep their culture and religion in mind but also to treat each patient like an individual, while respecting their decisions.

Case 3

Selma, 27 years old, Turkish descent, born in the Netherlands

My first pregnancy was planned. I had read a lot about pregnancies and was well prepared. My husband and I had already agreed to the prenatal screening. We both have a career and we decided to consider terminating the pregnancy if the child had Down syndrome or another disability, despite our religious and cultural background.

In my husband's family, there are a lot of children with birth defects. This is probably because in his family it is quite common to marry within the family. However, my husband and I are not related, so we knew we would have a smaller chance of conceiving a disabled child.

In our community a child with a disability is deemed a punishment of God. But actually, a child with Down's syndrome is a blessing, but not many people know this. Our religion actually states that children with a mental disability will go straight to heaven without being brought to account.

(Continued)

(Continued)

My midwife was very sweet. She assumed though, that I would not want a prenatal screening due to my background. She did not gave me information about the screening until I asked about it. I asked enough questions until I felt I could make a well-informed choice.

What I would like to make clear is that even though many women from certain cultural backgrounds do make their choices based upon social pressure, not all women do. Some women make their own choice. You, therefore, should not assume that all women from a certain cultural background will make the same choice.

Case 4

Aicha, 34 years old, Moroccan descent

I have two sons. Nouh is 6 years old and Said is 2 months old. Said was born with Down's syndrome. I found out that he had the syndrome after I gave birth to him. I had not used prenatal screening during my pregnancy. When we received that news, my husband and I were in complete shock. It was as if I had been thrown against a wall. After the initial shock, we quickly got ourselves back together and decided to do the best we could.

Our families were shocked as well but they lovingly supported us. Other people in our community did not know what to say. Sometimes they would ask rude questions like "Why didn't you do the prenatal screening test?," or "Will he always stay home with you?" I understand these kinds of questions are not ill intended, but as a parent I was not happy.

I was not prepared at all for a child with Down's syndrome but perhaps on a subconscious level, I felt it. I remember going to the playground with my oldest son while I was 7-month pregnant. I remember looking around at the other children and noticing there were not any children in our neighborhood with birth defects or disabilities.

When asked if I regret not taking the screening test, I answer "no." Even if I had known he had Down syndrome, I would not have done anything differently. I felt like someone's mom from the moment I was pregnant, despite what my child would turn out to be. To me having a "healthy" child is the same as having a challenged one. I do not believe it is a punishment from God, I believe it is something beautiful to receive a child.

Case 5

Ghaza, mother of four, Syrian descent

I emigrated to the Netherlands 4 years ago. My first three pregnancies were in Syria, and the fourth in the Netherlands.

In Syria, pregnancy is something you go through as a family. I left my family behind in Syria when I emigrated to the Netherlands. Not having my family around during the pregnancy made it difficult.

Healthcare in the Netherlands is much better than in Syria. I was a little disappointed that my midwife did not really understand my culture, but I do not blame her. When I went to the midwife, an interpreter was present because I do not speak Dutch fluently yet. The midwife asked if I would like prenatal screening and explained to me what that was. I declined. My religion teaches me that God has chosen my destiny and that of my children as well. I explained to the midwife why I declined, and she explained a little more about the screening but eventually respected my decision.

Best practice—tailored prenatal genetic counseling

Evidence-based strategies

This paragraph elaborates on some evidence-based strategies concerning themes to discuss with pregnant women with an immigrant background. Further, an exploration of communication pathways and the role of the interpreter during the provision of prenatal genetic counseling are provided. And finally, the possible supportive function of prenatal genetic health-care education is explained.

Themes to discuss in prenatal genetic counseling

In the provision of well-adjusted prenatal genetic counseling for pregnant women with an immigrant background, the following themes are important to discuss and to keep in mind (Kotsapas et al., 2009; Premkamur, 2009; Posch et al., 2012):

- religion beliefs and rituals;
- cultural beliefs and rituals;
- health-care utilization and assumptions;
- ethnicity-related diseases/syndromes; and
- consanguinity.

Communication

In the medical information provision to pregnant women with an inadequate language proficiency and health illiteracy, it is important to provide small amounts of information using simple language that meets the vocabulary of the women (Bennett et al., 2006; Boerleider et al., 2013b). It can also be helpful to practice the "teach-back method" which can contribute to improve patient satisfaction with the provided care (Centrella-Nigro and Alexander, 2017). It is a method that is based on the principle of asking the patient to restate information that has been presented to her. The "teach-back method" is an iterative process in which the genetic counselor is allowed to check for lapses in recall and understanding. It aims to reinforce and tailor messages, and engage the pregnant woman in an open dialog.

In general, in transcultural communication, the following points of attention are relevant (Middleton et al., 2005; Premkamur, 2009):

- invest in a trustful patient-health-care professional relationship;
- create continuity in health-care provision (location and counselor);
- respect the individuality of the pregnant woman;
- listen and watch carefully;
- use neutral language; and
- use an empathetic attitude.

The interpreter

Prenatal care providers experience the use of the professional interpreter as the most successful communication intervention to explore the health situation of the pregnant women who have a low language proficiency level. It is the most effective way to provide adequate care in terms of objective communication and to create the possibility to discuss sensitive and complex subjects (Kotsapas et al., 2009; Thomas et al., 2010; Boerleider et al., 2013a). The use of an interpreter coming from the social network of the pregnant woman is not recommended, because of the double role possibly played by this person. This type of interpreter can play the role of *interpreter, health-care professional, fellow visitor,* and second *patient* (van Rosse et al., 2016). All prenatal genetic counselors should be aware of situations in which a relative of the pregnant woman takes over the healthcare and content of information provision. In the paragraph, "practical instructions" advice to accomplish a cooperative relation with the translating relative is provided.

Health education

A good information provision about prenatal diagnostic testing combined with shared decision-making during prenatal genetic counseling is preconditional for making an informed decision about participation in prenatal diagnostics (Marteau et al., 2001; Michie et al., 2002). The passive provisions of leaflets to patients do not fulfill the needs of vulnerable patient groups and are outdated, unvalued, and ineffective (Young et al., 2017). Leaflets might be effective when they provide patients tailored information by using automated computer systems and adjusting the content of information on the knowledge and language proficiency level of the patient (Young et al., 2017). Research shows that for pregnant women with an immigrant background, especially the use of health illiteracy competent educational films and peer education provided by health educators, with the same cultural background and language of the attendees, can bridge language proficiency and health illiteracy barriers (Yu, 2011; Peters et al., 2014, 2017). These health educational strategies can contribute to increase knowledge about prenatal tests and healthcare and increase informed decision-making about participation in prenatal tests within pregnant women with an immigrant background (Peters et al., 2014, 2017). Unfortunately there are barely international examples of educational films about prenatal diagnostics. To our knowledge, none of them are adjusted for pregnant women with language proficiency problems and health illiteracy. A good example of an effective health educational film concerning first trimester screening on Down-, Edwards'- and Patau's syndrome is used in the Dutch prenatal screening program (Peters et al., 2017; National Institute for Public Health and the Environment Ministry of Health, 2020).

Practical instructions

Expecting mothers should be able to choose if they would like prenatal screening and further diagnostic examinations. Ideally the situation would be as follows: the midwife or gynecologist offers the pregnant woman sufficient and objective information so that the woman can make a well-informed choice. A woman is allowed to make her own choice, based on the objective information she receives and her personal values and customs. The medical professional supports the woman in her decision and helps her where needed.

Part of making an informed decision is the acquirement of sufficient knowledge. Keep in mind that a cultural background can play an important part in decision-making, but that the background is not equally important for everyone. Therefore each woman should be treated as an individual and should receive the same information.

Even though one's cultural background plays a part in the choices a person makes, research shows that ignorance plays a larger role in not choosing prenatal screening. Educated women from different backgrounds tend to choose prenatal screening. Educated people also tend to make healthier lifestyle choices. Armed with the right knowledge health-care professionals can help women make educated decisions. The right knowledge helps women resist emotional decisions and social pressure.

Sometimes, a physician becomes annoyed or upset when a woman does not want a prenatal screening, especially if the unborn child is at a higher risk of being born with a genetic disorder or birth defect. The annoyance of the physician results in a defensive attitude of the pregnant woman, with both parties not understanding each other and the woman losing trust in the physician. In this case the physician should show respect for the religious and cultural considerations of the expectant mother. He or she should inform her to his best ability and in all honesty but he or she should make clear to the parents that the choice is theirs and theirs alone. This will result in a more open conversation in which defensiveness is not needed, and this approach will be more constructive to both parties.

Language barrier

First, it is important for the physicians to assist the language proficiency of the patient. Sometimes patients seem to understand the information, but in reality do not. A way to test the language proficiency is to ask the patient to repeat the information the physician gave, but with his own words, or to ask the patient open questions. Based on the answers, the physician should be able to assess the comprehension of the patient. If the patient understands English well but not perfect, the physician should adjust the words he uses to the understanding of the patient and keep asking the patient to repeat the information to check if the patient understands correctly. Using visual information enhances the understanding of the patient, for instance a folder in the language of the patient, or images illustrating the medical problem or procedure.

If possible, if the patient does not understand English well enough, check if there are languages you both speak and understand that you can communicate in. A lot of Middle East countries also use French as second language, for instance. And many people speak Spanish as a second language. If the physician his or herself does not speak a second language, consider transferring the patient to a colleague who does.

Sometimes an interpreter is recommended in cases wherein the patient lacks proficiency in language skills. A nonprofessional interpreter may not understand all medical terms and, therefore, not convey the information correctly. Also using a relative as an interpreter can lead to the patient not being able to really open up about private matters. Another problem is that it is not good manners in certain cultures, in some Middle East countries, for example, to tell the patient bad news. It, therefore, is not uncommon that the family members withhold information from the patient so the patient does not become worried or stressed.

When an interpreter is needed, it is recommendable to use a professional interpreter when possible, especially when emotional or private matters are to be discussed. One should, however, look at each situation individually and evaluate the results of every setting. Is the information complex, that is, is it emotional information or are the medical terms and/or procedures difficult to follow? Does the physician think that the relationship between patient and nonprofessional interpreter will not come in the way of conveying information honestly? Is there a risk the informal interpreter will make mistakes in his translation, purposefully or by accident? Does the patient prefer an informal interpreter? The answers to these questions can help assess if a professional interpreter is needed or if an informal interpreter would suffice (Pharos, 2020).

Social network

In collective cultures the family is an important element. When treating a patient, the involvement of the family can help the patient. For elderly patients the children often play an important role in the care of the parents. For younger patients the parents and older siblings can help look out for the patient. Involving the family can help make sure the patient follows all medical advice and they can keep an extra eye out to make sure the patient is doing well.

When a patient is accompanied by people, ask them what their relationship to the patient is. Determine who you think is the most important in caring for the patient. Try to involve this person by also speaking to him or her and asking questions to make sure they too understand all information.

Self-assessment

Situation

Samira, 36 years old, emigrated from Morocco to the Netherlands 12 years ago

When I emigrated to Holland, I did not speak Dutch that well. I became pregnant with our first child shortly after coming to Holland. A relative made an appointment with the midwife for me and accompanied me to appointments. Due to the language barrier, my relative

(*Continued*)

(Continued)

spoke to the midwife. When they only spoke in Dutch, I could not follow the conversation and I felt left out.

It turned out the midwife noticed that my unborn baby had birth defects. This is probably genetic as I married my cousin. My relative understood what the midwife said but chose not to tell me; she thought the news would make me emotional and stress would complicate the pregnancy. I do understand why she did not tell me but would have preferred to have known. Of course I do not blame the midwife as she did not know what was going on.

Due to my own experience, my advice to patients with a language barrier is to choose for a professional interpreter instead of asking a relative to interpret. I also advise others to do their best to learn the language of the country they emigrated to as soon as they can. Whenever medical issues arise, the patient can explain himself and ask any questions he might have.

Questions:

1. Is there something the midwife could have done differently to make sure the woman understood the information given?
2. Would a professional interpreter be preferred in this situation? Explain your answer.
3. How could you involve a patient with a language barrier in the conversation when an interpreter is involved?
4. What could the midwife have done to try to communicate directly with the woman?

References

Agt H, Schoonen H, Fracheboud J, Mackenbach J: *Monitoring geinformeerde besluitvorming prenatale screening 2011, landelijke en regionale uitkomsten*, Bilthoven, 2012, National Institute for Public Health and the Environment.

Akhavan S: Midwives' views on factors that contribute to health care inequalities among immigrants in Sweden: a qualitative study, *Int J Equity Health* 11(47):2012.

Alcaniz L: *Waiting for bebé: a pregnancy guide for Latinas*, 2003, One World/Ballantine.

Alderliesten, M, T Vrijkotte, M van der Wal and G Bonsel (2007). Late start of antenatal care among ethnic minorities in a large cohort of pregnant women. BJOG 114: 1232-1239.

Aquino M, Edge D, Smith D: Pregnancy as an ideal time for intervention to address the complex needs of black and minority ethnic women: views of British midwives, *Midwifery* 31(3):373–379, 2015.

Atkinson D: Contrasting rhetorics/contrasting cultures: why contrastive rhetoric needs a better conceptualization of culture, *J Engl Acad Purp* 3:277–289, 2004.

Bastemeijer C, Voogt L, van Ewijk J, Hazelzet J: What do patient values and preferences mean? A taxonomy based on a systematic review of qualitative papers, *Patient Educ Couns* 100(5)(5):871–881, 2017.

Bennett I, Switzer J, Aguirre A, Evans K, Barg F: Breaking it down': patient-clinician communication and prenatal care among African women of low and high literacy, *Ann Fam Med* 4(4):334–340, 2006.

Boerleider A, Francke A, Manniën J, Wiegers T, Deville W: A mixture of positive a negative feelings: a qualitative study of primary care midwives' experiences with non-western clients living in the Netherlands, *Int J Nurs Stud* 50:1658–1666, 2013a.

Boerleider A, Manniën J, van Stenus C, Wiegers T, de Jong EF, Spelten E, Deville W: Explanatory factors for first and second-generation non-western women's inadequate prenatal care utilisation: a prospective cohort study, *BMC Pregnancy Childbirth* 98(15):2015.

Boerleider A, Wiegers T, Manniën J, Francke A, Deville W: Factors affecting the use of prenatal care by non-western women in industrialized western countries: a systematic review, *BMC Pregnancy Childbirth* 81(13):2013b.

Borges K, Vasilevska-Ristovska J, Hussain-Shamsy N, Patel V, Banh T, Hebert D, Pearl R, Radhakrishnan S, Piscione T, Licht C, Langlois V, Levin L, Strug L, Parekh R: Parental attitudes to genetic testing differ by ethnicity and immigration in childhood nephrotic syndrome: a cross-sectional study, *Can J Kidney Health Dis* 3(16):2016.

Browner C, Preloran H, Casado M, Bass H, Walker A: Genetic counseling gone awry: miscommunication between prenatal genetic service providers and Mexican-origin clients, *Soc Sci Med* 56(9):1933–1946, 2003.

Carlsson T, Balbas B, Mattsson E: Written narratives from immigrants following a prenatal diagnosis: qualitative exploratory study, *BMC Pregnancy Childbirth* 19:154, 2019.

Centrella-Nigro A, Alexander C: Using the teach-back method in patient education to improve patient satisfaction, *J Contin Educ Nurs* 48(1):47–52, 2017.

Cheng J, Guerra C, Pasick R, Schillinger D, Luce J, Joseph G: Cancer genetic counseling communication with low-income Chinese immigrants, *J Community Genet* 9(3):263–276, 2018.

Ford C, Harawa N: A new conceptualization of ethnicity for social epidemiologic and health equity research, *Soc Sci Med* 71(2):251–258, 2010.

Fransen M, Essink-Bot M, Vogel I, Mackenbach J, Steegers E, Wildschut H: Ethnic differences in informed decision-making about prenatal screening for Down's syndrome, *J Epidemiol Community Health* 63(3):262–268, 2010.

Gal I, Prigat A: Why organizations continue to create patient information leaflets with readability and usability problems: an exploratory study, *Health Educ Res* 20(4):485–493, 2005.

Garza G, Hodges-Delgado P, Hoskovec J, Palos G, Wagner C, Zacharias N, Noblin S: Exploring experiences and expectations of prenatal health care and genetic counseling/testing in immigrant Latinas, *J Genet Couns* 29(4):530–541, 2020.

Geertz C: *The interpretation of cultures*, New York, 1973, Basic Books, Inc., Publishers.

Gitsels J, Manniën J, Gitsels L, Reinders H, Verhoeven P, Ghaly M, Klomp T, Hutton E: Prenatal screening for congenital anomalies: exploring midwives' perceptions of counseling clients with religious backgrounds, *BMC Pregnancy Childbirth* 14:237, 2014a.

Gitsels J, Manniën J, Ghaly M, Verhoeven P, Hutton E, Reinders H: The role of religion in decision-making on antenatal screening of congenital anomalies: a qualitative study amongst Muslim Turkish origin immigrants, *Midwifery* 30(3):297–302, 2014b.

Gitsels J, Verhoeven P, Manniën J, Martin L, Reinders H, Spelten E, Hutton E: Factors affecting the uptake of prenatal screening tests for congenital anomalies; a multicentre prospective cohort study, *BMC Pregnancy Childbirth* 9(14):264, 2014.

Heetkamp KM, Peters I, Bertens L, Knapen M: An unwanted pregnancy and language proficiency level are associated with first antenatal visit after the first trimester: results from a prospective cohort study, *Midwifery* 89:2020.

Hofstede G: *Allemaal andersdenkenden, omgaan met cultuurverschillen*, Amsterdam, 2010, Uitgeverij Contact.

Hofstede_Insights: *Compare countries*, 2020. From https://www.hofstede-insights.com/product/compare-countries/.

Jun M, Thongpriwan V, Choi J, Choi KS, Anderson G: Decision-making about prenatal genetic testing among pregnant Korean-American women, *Midwifery* 56:128–134, 2018.

Kalengayi FN, Hurtig A, Ahlm C, Ahlberg B: "It is a challenge to do it the right way": an interpretive description of caregivers' experiences in caring for migrant patients in Northern Sweden, *BMC Health Serv Res* (12):433, 2012.

Kathir I: *Tafsir of Ibn Kathir*, Riyadh, 2003, Darussalam Global Leader in Islam Books.

Khadeejah A: *Abortion and its ruling in Islam (Ethics 1.4)*, 2020. From https://www.abukhadeejah.com/abortion-and-its-ruling-in-islam-redstone-academy-ethics-1-4/.

Kotsapas C, Dixon C, Nauta M, Fakokunde A, Yoong W: Transcultural considerations in obstetrics and gynaecology: what the clinician needs to know, *J Obstet Gynecol* 29(3):2009.

Macionis J, Plummer K: *Sociology. A Global Introduction*, 2011, Pearson Education Limited.

Marteau T, Dormandy E, Michie S: A measure of informed choice, *Health Expect* 4(2):99–108, 2001.

Martin L, van Dulmen S, Spelten E, De Jonge A, De Cock P, Hutton E: Prenatal counseling for congenital anomaly tests: parental preferences and perceptions of midwife performance, *Prenat Diagn* 33:1–12, 2013.

Michie S, Dormandy E, Marteau TM: The multi-dimensional measure of informed choice: a validation study, *Patient Educ Couns* 48(1):87–91, 2002.

Middleton A, Ahmed M, Levene S: Tailoring genetic information and services to clients culture, knowledge and language level, *Nurs Stand* 20(2):52–56, 2005.

National Institute for Public Health and the Environment Ministry of Health, Welfare and Sport. Screening. Link on webpage 'Dutch Prenatal Screening program' to cultural competent film, 2020. From https://vimeo.com/249632107.

Peters I, Posthumus A, Steegers E, Denktaş S: Satisfaction with obstetric care in a population of low-educated native Dutch and nonwestern minority women. Focus group research, *PLoS One* 14(1): e0210506, 2019.

Peters I, Posthumus A, Reijerink-Verheij J, Agt HV, Knapen M, Denktaş S: Effect of culturally competent educational films about prenatal screening on informed decision making of pregnant women in the Netherlands, *Patient Educ Couns* 100(4):776–782, 2017.

Peters I, Schölmerich V, van Veen D, Steegers E, Denktaş S: Reproductive health peer education for multicultural target groups, *J Multicult Educ* 8(3):162–178, 2014.

Pharos: *Kwaliteitsnorm tolkgebruik bij anderstaligen in de zorg*, 2020. From https://www.pharos.nl/kennisbank/kwaliteitsnorm-tolkgebruik-bij-anderstaligen/.

Pimentel V, Eckardt M: More than interpreters needed: the specialized care of the immigrant pregnant patient, *Obstet Gynecol Surv* 69:490–500, 2014.

Poeran J, Maas A, Birnie E, Denktas S, Steegers E, Bonsel G: Social deprivation and adverse perinatal outcomes among Western and non-Western pregnant women in a Dutch urban population, *Soc Sci Med* 83:42–49, 2013.

Posch A, Springer S, Langer M, Blaicher W, Streubel B, Schmid M: Prenatal genetic counseling and consanguinity, *Prenat Diagn* 32(12):1–6, 2012.

Posthumus A, Borsboom G, Poeran J, Steegers S, Bonsel G: Geographical, ethnic and socio-economic differences in utilization of obstetric care in the Netherlands, *PLoS One* 11(6):2016.

Premkamur G: Are we good at managing pregnancy in ethnic minority? *J Obstet Gynaecol* 28(4): 373–376, 2009.

Schouten B, Meeuwesen L: Cultural differences in medical communication: a review of the literature, *Patient Educ Couns* 64(1–3):21–34, 2006.

Telushkin J: *Jewish Literacy: the most important things to know about the Jewish religion, its people, and its history*, New York, 1991, HarperCollins.

The Jewish Encyclopedia: *The unedited full-text of the 1906 Jewish Encyclopedia*, 2020. From http://www.jewishencyclopedia.com/articles/9028-judaism#.

Thomas P, Beckmann M, Gibbons K: The effect of cultural and linguistic diversity on pregnancy outcome, *Aust N Z J Obstet Gynaecol* 50:419–422, 2010.

Tschudin S, Huang D, Mor-Gültekin H, Alder J, Bitzer J, Tercanli S: Prenatal counseling—implications of the cultural background of pregnant women on information processing, emotional response and acceptance, *Ultraschall Med* 32(Suppl 2):E100–E107, 2011.

van Rosse F, Suurmond J, Wagner C, de Bruijne M, Essink-Bot M: Role of relatives of ethnic minority patients in patient safety in hospital care: a qualitative study, *BMJ Open* 6(e009052):2016.

Wiegers T, Deville W, de Jager M, Plass A: *De (niet-)geïnformeerde keuze van zwangere vrouwen van Turkse en Marokkaanse afkomst, zwangere vrouwen met lage SES en jonge zwangere vrouwen, rondom deelname aan prenatale screening op downsyndroom en het structureel echoscopisch onderzoek (SEO)*, 2015, NIVEL.

Williams R: *A vocabulary of culture and society*, New York, 1983, Oxford University Press.

Young A, Tordoff J, Smith A: What do patients want?' Tailoring medicines information to meet patients' needs, *Res Soc Adm Pharm* 13(6):1186–1190, 2017.

Yu J: A systematic review of issues around antenatal screening and prenatal diagnostic testing for genetic disorders: women of Asian origin in western countries, *Health Soc Care Community* 20(4):329–346, 2011.

Zagefka H: The concept of ethnicity in social psychological research: definitional issues, *Int J Intercul Relat* 33:228–241, 2009.

Section II Patient journey and offering support

Introduction

The decision to terminate a desired pregnancy, which in many cases is on account of a non-lethal fetal congenital abnormality, is an emotionally overwhelming and complex process for prospective couples (McCoyd, 2010; Korenromp et al., 2009) and to most the most difficult decision of their lifetime (Hodgson et al., 2016). Pregnant couples rarely anticipate having to make an active, voluntary decision concerning a termination of their pregnancy (Geerinck-Vercammen and Kanhai, 2003). Although supported by health care professionals (HCP'S), the burden of the decision whether to terminate or continue the pregnancy lies with the couples (Korenromp, 2006). HCP's have are tasked with helping couples in every possible way to make this life changing decision (Hodgson and McClaren, 2018); providing care, decision-making support (Hawkins et al., 2013) and aftercare. This section offers analyses, advice and insight to HCP's working in prenatal genetic counseling or encountering couples faced with prenatal anomalies.

Reading this entire section, provides the reader with a wealth of insights, mostly of quite practical nature. In addition to its scientific underpinnings, all the authors have written about their experiences straight from the heart. This is not a coincidence. Our experience has led to the conviction that the best care for pregnant couples faced with fetal anomalies comes straight from the heart (Ramdaney et al., 2015).

At the heart of this section are the narratives of Lara mother of Mae, and Iris, mother of Noud. Couples in prenatal genetics generally describe their experiences according to distinguishable phases, and therefore the chapters in this section follows these phases: the discovery, the decision-making process, giving birth, grief and building a life again. A variety of authors reflect on the narratives of Lara and Iris from their professional experiences: three psychologists working with pregnant couples in prenatal genetics within the context of an academic hospital, Sam Riedijk (editor) and Frederike Dekkers and Iris Jansen, two gynecologists, Eva Pajkrt and Liesbeth van Leeuwen (VUMC), the director of the UK charity Antenatal Results and Choices (ARC) Jane Fisher, a number of third year students reflecting on their education session with Lara and Iris, Diana Zavyalova and Jasmijn Klapwijk, minor student and PhD student in prenatal genetics respectively, and Karin Diderich (editor).

As this section is heavy with Dutch authors, many of the recounts stem from the Dutch context in which pregnancy ending is legally allowed until 24 weeks gestation, in most cases by induced labor. As authors we would like to emphasize that we have no intention to be prescriptive. The experiences and reflections in this section happen to be situated within this context, however we do not deem this context to better or worse than other contexts. As HCP's we feel the responsibility to do everything within our power to provide adequate (after) care and support couples in making decisions within the prenatal genetics context that are best fitting for them. Best fitting decision are decisions that in hindsight remain decisions that couples and individuals may carry, without regret. We hope this section will contribute to a thorough understanding of the experiences and needs of couples in the context of prenatal genetics.

References

Geerinck-Vercammen CR, Kanhai HHH: Coping with termination of pregnancy for fetal abnormality in a supportive environment, *Prenatal Diagnosis* 23:543–548, 2003.

Hawkins A, Stenzel A, Taylor J, Chock VY, Hudgins L: Variables influencing pregnancy termination following prenatal diagnosis of fetal chromosome abnormalities, *J Genet Couns* 22:238–248, 2013.

Hodgson J, McClaren BJ: Parental experiences after prenatal diagnosis of fetal abnormality, *Semin Fetal Neonatal Med* 23:150–154, 2018.

Hodgson J, Pitt P, Metcalfe S, et al: Experiences of prenatal diagnosis and decision-making about termination of pregnancy: A qualitative study, *Aust N Z J Obstet Gynaecol* 56:605–613, 2016.

Korenromp M, Page-Christiaens GCML, van den Bout J, Mulder EJH, Visser GHA: Adjustment to termination of pregnancy for fetal anomaly: a longitudinal study in women at 4, 8, and 16 months, *Am J Obstet Gynecol* 201(160):e1–e7, 2009.

Korenromp M: *Parental adaptation to termination of pregnancy for fetal anomalies*, Utrecht, 2006, Universiteit Utrecht.

McCoyd JL: Authoritative knowledge, the technological imperative and women's responses to prenatal diagnostic technologies, *Cult Med Psychiatry* 34:590–614, 2010.

Ramdaney A, Hashmi SS, Monga M, Carter R, Czerwinski J: Support Desired by Women Following Termination of Pregnancy for a Fetal Anomaly, *J Genet Couns* 24:952–960, 2015.

CHAPTER 6

The discovery

Frederike Dekkers[1], Lara Ras[2], Iris Emons[2], Sam Riedijk[3], Jane Fisher[4] and Karin E.M. Diderich[3]
[1]Psychotherapy, Private Practice, Rotterdam, The Netherlands
[2]Experience expert, The Netherlands
[3]Department of Clinical Genetics, Erasmus Medical Centre, Rotterdam, The Netherlands
[4]Antenatal Results and Choices (ARC), London, United Kingdom

Both Lara and Iris are Dutch. In the Netherlands, couples are legally allowed to terminate a pregnancy for fetal congenital abnormalities up to 24 weeks of pregnancy.

Lara: Mae

Hugo and I had been in a relationship for 10 years when we decided that we very much wanted to become parents. During the summer holidays, I stopped my birth control and we started waiting to get pregnant. The first time we were very happy. But the first ultrasound at 10-week gestation showed that the fetus had already stopped growing around 6-week gestation. I was in shock, and the thought of a lifeless fetus in my body was awful. I wanted it out of my body as fast as possible. After this episode, I could see what happened in perspective; something was wrong with the fetus, and it was a pity, but Mother Nature had done her work properly and had made the decision to end this pregnancy for us. After this miscarriage, it took us a long time to get pregnant again. However, as soon as I was pregnant again I started to feel intensely "pregnant" and a little uncertain.

I was hyperconscious of the fact that being pregnant did not equal having a baby. To reassure myself, I requested an ultrasound already at 7-week gestation. Everything looked fine, but I was not reassured. Because of this uncertainty, I waited a long time before informing people around us. I waited the longest with informing the teachers and students of the high school where I was working as a teacher.

Nevertheless, the pregnancy continued and the 20-week ultrasound was nearing; there was no one available at 20 weeks, so I had mine at 21-week gestation.

I remember that I had noticed some pupils in school staring at my belly and I decided that indeed "maybe it is time to share this news with the world." We were just like other people who have never lost a child and were mostly curious to learn whether we would have a boy or a girl.

The sonographer who performed the ultrasound was very positive. She explained everything she saw and lectured us on brain development and on the intestines of the

Prenatal Genetic Counseling
DOI: https://doi.org/10.1016/B978-0-12-823329-0.00011-8

baby and she was pointing out all kinds of interesting aspects of the baby on the screen. However, during the ultrasound I also noticed that she was printing a lot of pictures. I did not think much of it and assumed that I would be given many pictures to bring home. She could not tell whether our baby was a boy or a girl, because it was not well positioned and refused to turn over. I felt increasingly relaxed with all this good news. As I was rubbing the gel off my belly, the sonographer told me she had seen something remarkable about the thigh bone of our baby. It had a prominent angle and other than being noticeable she had no idea whether it would be something severe or not, so she referred us to her supervisor with urgency. She called him and asked whether we could come over that same evening. And with this news, we were outside. In total panic...

I remember feeling intensely betrayed and I could not grasp that during the entire ultrasound, the sonographer had reassured us that everything was all right, only to crush it at the very end of the session. We would have so much rather wanted to learn as soon as she did that something was amiss.

[Years later I was informed that this sonographer was following protocol. The reason behind this was that pregnant couples are so shocked by bad news that they will not be able to hear anything else, including all those aspects that are good about the baby. It is my personal opinion that a couple is much better supported if the midwife would tell the bad news right away and ask the couple what they need. Something like: "What do you need now?," "Are you able to digest what I am telling you?," and "Shall we continue this ultrasound at a later moment in time?" would have worked for us. Also, it is my impression that sonographers struggle with telling bad news, which I do understand. However, I cannot imagine that many couples would blame the sonographer for it. Yes, people will be emotional and scared for what they are faced with, but that is not the sonographer's fault.]

That same evening Hugo and me went for a more extensive ultrasound together with my mother. A new 2D and 3D ultrasound were made and soon it showed the presence of many more anomalies in our baby. The other thigh bone was also "hooked" and the baby had two clubfeet. At first we hoped that the baby would merely have an orthopedic problem, which could be taken care of after birth. But our hope started to diminish. I can remember quite well that the man who made the ultrasound asked me whether he could inspect the baby's face with 3D ultrasound. Back then in 2011 the use of 3D was not as widespread as it is now and it felt like a present. Yet, I really did not want it. It was at that moment, laying down on that table that the thought hit me that something was terribly wrong with my baby. Mentally, I felt something changed. I was distancing myself from my pregnancy and my baby. It seemed awful to me to see the face of the baby, and it would make everything much too real. We did find out that our baby was most likely a daughter.

After that terrible news, we were referred to a fetal medicine unit in a tertiary care hospital and I went back to work. I guess I was hoping this would allow me to ignore everything that was going on, which of course did not prove to be very successful. Instead, I only informed my supervisor and returned home.

In the evening, we were told that the first available time slot for further investigation of the baby was after the weekend. And that was the start of the true roller coaster.

The investigation started with another 2D ultrasound and a meeting with a clinical geneticist. The next day, we needed to draw blood and had another 3D ultrasound. We were surprised that the clinical geneticist took part in the ultrasound session. She indicated that she wanted to see for herself and she wanted to have the opportunity to ask zoom in on certain details. The fact that she joined in strengthened me. I got the feeling that she shared our sense of urgency to clarify the anomalies, and that she was determined to do everything within her power to understand what was wrong with our baby. It feels good when an expert asks additional questions. I felt that me, as a lay person, could ask the right questions, but the expert would ensure that no details would be overlooked.

During the 3D ultrasound, many more anomalies were found in our baby. Her face looked different than would be expected based on the looks of me and Hugo. Her nose was very flat and wide and the ears were set much lower than is normal. The umbilical cord had one instead of two arteries. It also became apparent that her ribs were not round but crooked and were much too close to each other, posing the threat that her lungs could not properly develop due to the lack of space. The sonographer was in doubt about the sex of the baby.

The day after the ultrasound, we had an amniocentesis. We had not properly thought this through so we went to the hospital on our bicycles. I was very nervous and worried about experiencing pain form the puncture. The doctor performing the puncture aligned well with me and taught me some tricks to get through the procedure, for instance, by wiggling my toes. It worked well for me. The fact that he asked me whether I was nervous and gave me this advice made me feel acknowledged. That indeed this was a frightening and unpleasant procedure and that my tension was valid. After the puncture, we rested a little in a special room and then turned back home on our bicycles.

After all these investigations the long waiting started for the results of the ultrasound scans, the chromosomal microarray, and the DNA tests.

We only had our 20-week ultrasound at 21-week gestation and because of that I felt intensely time pressured. In our country, we are legally allowed to end a pregnancy until 24-week gestation and by now I was 22-week pregnant; I felt the baby move and my belly became more visible each day. I felt terrible about having missed an entire week in which more diagnosis could have taken place.

It was difficult for me to position myself in relation to the baby. We did not know whether we could continue this pregnancy and I felt the need to keep some distance to the child but the big belly and baby movements were impossible to ignore.

Frederike Dekkers

Developments in prenatal screening, prenatal ultrasound, and genetic testing have enabled the detection of a growing range of fetal congenital abnormality and genetic conditions (Minear et al., 2015; Hui and Bianchi, 2017; de Wit et al., 2014; Best et al., 2018). Consequently, increasing numbers of women and their partners are confronted with the difficult decision of whether to continue with or terminate a pregnancy. And right there, the emotional and psychological challenge for couples begins. Studies show that women are unprepared for the level and duration of the emotional pain and the "roller coaster" of emotions experienced during and after a termination of pregnancy (Lafarge et al., 2014; Jones et al., 2017).

Psychological challenges during the phase of diagnosis in decision-making

How do we stand our ground, when our perception of the future falls apart and, simultaneously, we have to understand difficult information to decide on further diagnostic evaluation?

The moment couples express their wish for a child with each other, a new perception of the future unfolds in their minds. They start to think about themselves as parents, and they reflect on their own parents. Couples see themselves in a nursery room, walking behind a stroller, and they fantasize about a son or a daughter to cuddle, care for, and play with. This is a natural and automatic process that enables couples to prepare themselves for the big upcoming life-changing event called parenthood. If congenital abnormalities are found through prenatal testing, this naturally developing process comes to a halt, shattering the mental image that couples have been forming (Lou et al., 2020).

It is our duty as health-care professionals (HCPs) to help couples handle this remorseless reality and guide them gently from fantasizing about the future to understanding the often complicated and specific information about the congenital abnormalities found. It is our task to help them understand all the information and all the possible next steps to take as well as to assist them in coping with this "earthquake" and all associated fears.

In light of a cross-sectional study, we asked couples what kind of support was most meaningful to them and what they missed mostly (Dekkers et al., 2019). One of the main needs couples expressed is time. Time to ask all their questions to an HCP with specific knowledge, this will mostly be the fetal medicine specialist, but sometimes other professionals can also be involved such as a pediatric cardiologist or neurologist, or clinical geneticist. They need someone who can provide detailed information and help them to answer questions like "What are we facing?"; "Which techniques are available and what are the benefits and risks?"; and "How will that help us?."

Help couples understand the information

When we humans are under a lot of stress and pressure, our minds and memory typically stop functioning optimally (Sandi, 2013). This entails that couples may need to repeatedly hear the same information to be able to thoroughly comprehend it, maybe even during consecutive sessions, which asks of HCPs to avoid the use of too many medical terms. Having that opportunity, whether it is via a repeated appointment or contact by telephone or e-mail, is essential. If available, written information about the congenital abnormalities, or a reference to reliable information online, also helps to guide couples in their decision-making process for further diagnostic evaluation (Skirton and Patch, 2013). Similarly, reliable information helps couples in discussing the findings and the process they are going through with their family and friends.

Couples report that they feel reassured if an HCP makes time, shows his/her availability, and knows their specific file well (Dekkers et al., 2019). Couples say that they prefer facts and HCPs being honest about not yet knowing, rather than trying to "mask" possible negative outcomes.

Couples reflected that, during this timeframe, it was not always clear to them what type of appointment with the HCPs to expect. Bad news sometimes came as a surprise, when only partial test results were anticipated.

When couples are referred to a tertiary care center for further diagnostic evaluation, couples reflect that they are not always prepared by HCPs for which kind of evaluation or tests they may have to go through. Some couples were referred for an ultrasound but had to suddenly decide whether they wanted to do an amniocentesis. Some couples are glad this can be done right away, but some couples reflect it was hard to decide then and there; not only aware of the need for the information an amniocentesis produces, but also very aware of the miscarriage risks involved. In a period of great insecurity, it is key to ensure that couples know the nature of an appointment, so that they can prepare. Couples feel lost and overwhelmed, and being able to prepare gives them a sense of control.

Understandably, couples appreciate quickness in the sequence of appointments, keeping the time spent waiting for possible bad news to a minimum. And they prefer one "key" HCP, who guides them throughout the whole period and gets to know them. This can also help to avoid contradictory or incomplete information.

For couples, this situation is an unknown area and they fully depend on the sensitivity, responsiveness, and behavior of the HCPs they encounter. Often the couples in this situation have to share the waiting room with mostly happy couples. It helps couples if the HCP acknowledges this and gives words to the burden this represents to the couples. Couples appreciate humanity, empathy, acknowledgment, and understanding for the exceptionally difficult task they are facing and the roller coaster of emotions that comes with it.

Case A

This case concerns a couple's first pregnancy. The 21-week ultrasound showed a lack of target signs and a dilation of the intestine, which caused a strong suspicion of an anal atresia. The father of the mother had just passed away after colon cancer, living with a stoma his last years. Parents very much feared their child living in need of a stoma. Mother's father had suffered from many complications.

Parents went to different specialists for a second opinion, also in another city. It was a very stressful period because they were afraid of making the wrong decision. They felt the need to talk to as many specialists as possible and gather as much information as they could to ascertain the severity of the anomaly. But seeing so many specialists did not lead to a clearer picture of the anomaly. Some specialists were outspoken, while others were not. In the end, the parents were left with the same question: how severely affected is our child and what should we do with this pregnancy? They felt time ticking away. Finally, they came back to the first HCP and decided to terminate the pregnancy. At birth, it became clear that it was indeed a severe form of anal atresia. The parents felt relieved, seeing with their own eyes how severe the anomaly was. Having visible "proof" helped them to feel that they had made the right decision. They were very grateful for all the help they received from all the HCPs.

Not all HCPs understood why the couple consulted so many specialists. Although all the support of the HCPs did not provide the parents with the clarity they sought, they did feel that they had done everything in their power to get it. Therefore they were satisfied with the manner in which they handled the situation.

What can we learn?

When the severity of the detected anomaly is hard to diagnose, it is extremely difficult for parents to decide about the life of their child. In this diagnostic evaluation phase, parents seek clarity—facts and numbers on which to base their decision. Consulting multiple HCPs does not always provide clarity. Having to integrate all the information is stressful and is not something that all parents can cope with. The parents who can have expressed that this helps them to feel that they did everything in their power to make the best possible decision. It is up to us as HCPs to help parents oversee all the options, fathom the implications of these options, and select which option suits parents best. It is also our responsibility to accept that parents follow their own path to reach a decision. After all, it is the parents who have to live with their decision for the rest of their lives.

This case also shows how previous experiences with a specific disease or anomaly can impact the way parents think about the quality of life for their child and how they can cope with that.

Iris: Noud

I was 33 years old, had a healthy 2.5-year-old son, and I just really wanted to have a second child. That was all I wanted. It seems so simple, but unfortunately it turned my whole life upside down for ever.

I got pregnant quickly, just like my first pregnancy, and was unaware of the possibility that something could go wrong when I was first visiting the midwife at 11-week gestation.

I noticed immediately on the routine ultrasound that the baby seemed to be folded over, sort of double, and seemed not to have enough space. According to the midwife, this was normal, nothing to worry about. A week later I was at the hospital for a nuchal translucency measurement. This time we were tense because this test could signal that something was amiss. Again, I saw the baby folded over on the ultrasound. When I told the sonographer that last time the baby was in the same position, she left to confer with a gynecologist. There seemed to be a lack of amniotic fluid, which was an indication of a kidney problem. The NT measurement was within normal range though.

For others, it is difficult to imagine how you feel when you are told something like that. It seems unreal, are they really saying that something is fundamentally wrong, or is there something wrong that may be repaired? You feel unable to understand, and to get your head around it. It was decided that more investigation was needed, and because of a national holiday, we could only come back to the hospital later during that week. The doctor made a few phone calls, and it turned out we could go for an extensive ultrasound scan faster if we were to go to a different hospital, so we did.

That ultrasound confirmed a lack of amniotic fluid. Because it was the national holiday, a residential doctor was on duty and she could only say that it did not look good. A pregnancy with this little amniotic fluid would be difficult but exactly how difficult was hard to tell. She told us that if we were to end the pregnancy, we would need to go into labor because it was overdue for a surgical abortion.

I remember well because until then all I thought was "If it turns out to be wrong they should probably take it away." But I had never assumed that I would have to go into labor. We walked out from that hospital feeling drained and we could not believe this was happening to us, as if it were happening to somebody else. What is going on here? Is this real?

We had another appointment with a gynecologist 3 days later. We expected him to give us the confirmation that this pregnancy was doomed. Here I was again, on my back on the table with yet another gynecologist, but still at hospital number 2. During the ultrasound a lot of amniotic fluid was visible, and the baby was moving up and down, just like my son Mats did when I was pregnant with him. The gynecologist said that everything looked good, there was enough amniotic fluid and "look, you see his little hand moving here with little fingers." I will never forget that he said that. He said, "Look, he seems to be waiving at you with his hand and here you can see his little fingers, look." However, we still needed to wait until the 20-week ultrasound scan to get the certainty that everything was really all right. We left the room with a referral for a 20-week ultrasound scan 5 weeks later, at 18-week gestation. I lost some blood and, therefore, I was informed that I had to slow down and I was not allowed to take a bath.

We left feeling ambiguous. We felt that something was wrong and we also felt relieved that all might be okay. We could not entirely believe it. I remember a friend sending me a text message saying that "sometimes miracles may truly happen."

I regularly lost blood and each time I went to the loo I was afraid to find blood in my underwear. People around us told us that we should try to have some faith because when the doctor says that all is fine, you can believe it. Others said that they too lost some blood during pregnancy and that they also had given birth to a healthy baby.

And so the weeks passed and we started to wonder whether it was a boy or a girl. And maybe everything would be okay after all, because the baby was still growing and moving in my belly. I saw some fabulous tapestry with little elephants, and I thought if it is a boy then it would be great to have this tapestry on the wall in his room. I kept working all this time because there I could just be me; nobody knew what was going on. I would have a few hours to breathe and forget about my preoccupation with this pregnancy.

Then finally the day arrived that we had to go to hospital number 3, the Erasmus Medical Center for our appointment. I had to register and have my picture taken for a pass. Until this day I have held on to this pass, because it reminds me of the time when I still had hope.

We were sitting in the waiting room and you realize that all these pregnant couples are here for a reason. Some people seemed relaxed, others tense, and others drained.

We were called in by the sonographer, and that was the moment we had been anticipating all these weeks. I was laying there and was so nervous about there being enough amniotic fluid. That was the only thing I had been thinking about for weeks. Is there enough fluid and are the kidneys working well? She immediately started scanning and she saw a lot of fluid and the baby was bouncing, moving and seemed to be at ease. She said straight away that the amniotic fluid seemed all right and that she was looking at two kidneys that seemed to be in order too. She continued the scan and said that the heart, the brain, and the other organs also looked all fine. She asked us whether we would like to know if it is a boy or a girl? And of course we wanted to know! I got quite enthusiastic, this was going really well, I could hardly believe it. She said "It's a boy."

A boy, I would be the mother of two little boys. That was amazing, two of a kind, what I always dreamed of. I did not care about the sex but I liked two of the same sex because that would be most fun for them. At that moment, I started to think about the elephant tapestry again and I was just dreaming about that a little bit, drifting off in the new room with the nice elephant tapestry. Then, while I was still in that day dream, she said: "I see an anomaly around his mouth, it seems to be a cleft lip." I was thinking, wait what did she say? Cleft lip? "You said it looks like a cleft lip? Is it or does it just look like it?." "No I am sure there is a cleft lip," she said. That was a blow. I was thinking he will have a cleft lip, while his brother is super handsome with blond hair and blue eyes, but okay, we will deal with that. But then she said, "I see a cleft on both sides and I also see an anomalous arm." And I thought no, that is not possible, because doctor 2 told that the baby was waiving at us. There could not possibly be something wrong with his arm. But she keeps going and says that he also seems to miss some fingers, and it seems that one arm is shorter than the other one and also has a strange position. And I thought no, he has fingers and hands, and now you tell me this, that is just not right. She proceeds by telling us that often when there are multiple anomalies, we call this a syndrome. She advises us to opt for an amniocentesis. And then she lost me. I was in the room and all of a sudden all the hope and faith I had for this pregnancy vanished. The roller coaster of all the emotions I had experienced these last weeks became more than I could bear and I broke down. I could not take in anymore.

What helped Iris:

- Be clear, there is no way to make bad news good. Be direct and do not circle around the bad news.
- Remember that all you will say will be remembered and chewed upon. The only grip a pregnant couple can experience are the words of the expert, so be mindful of your words.
- A fixed person who speaks your language and guides the pregnant couple through these taxing times.

What did not help Iris:

- Holding on to bad news until the end. I would have preferred being informed the moment she saw on the ultrasound.
- I was in three different hospitals and spoke to a different expert each time. This makes for unclear communication and caused a lot of distress. Who should you believe and put your faith in?

Sam Riedijk

Since the implementation of the 20-week ultrasound screening program in the Netherlands, I have been working as a medical psychologist within the prenatal genetics subdivision of the Clinical Genetics Department. The desire to share our lessons learned is the motive behind this book.

Between 2007 and nowadays, I have met many pregnant couples that have made decisions about continuing or ending a pregnancy based on fetal anomalies. This wealth of clinical experience has led to a narrative of the "average experience" of couples with decision-making, ending a pregnancy, and processing grief in the short as well as the longer term that I often find myself reciting during counseling sessions. The provision of a frame of reference in the form of other couples' experiences is often viewed by couples as the most valuable aspect of these counseling sessions. Not only do couples indicate to recognize their own processes, but they also like to be able to contrast themselves, see how they differ, to get grip on what is happening to them.

Roller coaster

Most couples encounter the fetal anomalies during the 20-week ultrasound screening. To most, this comes as a devastating surprise. Also, this is the moment that psychological responses between pregnant women and their partners already differ significantly (Kaasen et al., 2010). Many couples describe that while knowing the screening is not just intended to discover the baby's sex, they could not fathom that something being wrong could happen to them (Hammond et al., 2020).

After the initial discovery, couples are referred for advanced ultrasound scanning (United States) in an academic or specialized clinic. This is where the anomaly can be confirmed and inspected in more detail. Despite being very much aware of the reason for referral, couples describe that in the United States, for each test result, they also cherish hope that the anomaly is not as bad or is treatable at least. If in fact the anomaly is worse or has unknown or poorer prognosis, couples describe falling deeper. Most couples have used the term "roller coaster" to describe this diagnostic phase. The roller coaster experience is caused by feelings of despair and hope, being overwhelmed with information and having to speak with many HCPs in a short time.

Couples need to have organizational skills to keep grip on their schedule. The strong emotions couples experience in this period exhaust them, as well as interfere with information processing (Sandi, 2013). To understand the cause and nature of the anomaly, couples may be referred to a variety of medical specialists, including clinical geneticists. Often, the fetal medicine specialist, or clinical geneticist, keeps the overview and pieces together the information that couples gather during their various encounters with experts. Couples crave an HCP who will actively be involved in piecing together the picture and who is readily available and straightforward about what is known, unknown, or still unclear (Dekkers et al., 2019). In my experience, couples highly appreciate HCPs who are transparent about what meaning they attribute to the findings. An HCP stating: "This looks bad" provides much needed interpretation. For couples, it is frightening to have to rely on their own interpretation, since this is their child, and they are not medical experts. And even if normally they are experts, when it comes to their own pregnancy, and their own child, they are just as "lay" as anyone else.

Nondirectiveness is the needed counseling approach when it comes to supporting decision-making (Evans, 2006). But not when it comes to making sense of ultrasound findings, test results, and the overall picture of the condition and prognosis of the unborn child (Sullivan et al., 2019). Here the HCP has to medically "color the painting." "From a medical perspective, this is worrying. Your child will need treatment and operations." It is only after this that couples can apply their personal meaning to the medical picture. "Our child will need treatment and operations, and we are highly motivated to go for it" or "Our child will need treatment and operations, and we feel that we do not want to put this burden on our child."

Jane Fisher

Through Antenatal Results and Choices—ARC's history, prenatal testing technologies have changed dramatically and there has been an exponential increase in prenatal diagnoses (Lewis et al., 2014). What has remained a constant is the psychological impact on the woman and couple who are given the news that there is a suspected or confirmed fetal anomaly in the context of a wanted pregnancy. Much has been done by the NHS to improve pretest information, but, however, good this is, it will do little to psychologically prepare most expectant parents for unexpected or difficult news in the pregnancy.

> *I was knocked right off cloud nine and it finally dawned on me that this was the only reason we were here — to find out if our baby had a major abnormality. The rest of the time in that room went by in a blur. I have only a vague recollection of what happened next. I remember hearing the words spina bifida.*
>
> ***ARC News, August 2014.***

Prenatal diagnosis of fetal anomaly requires parents to confront the loss of the "healthy" baby they had previously conceptualized and built their hopes and expectations around. Some need time to accept the reality of the diagnosis and its possible implications. It can be difficult for them to adjust their expectations and begin to envisage a different future. Some feel the need to have the anomaly confirmed by a second opinion, not because the initial clinical judgment is wrong or that they doubt the medical facts, but because hearing the same information from another source is part of the process of accepting the reality of the situation and understanding what it may mean for them (Aite et al., 2011). Most diagnoses, apart from fatal conditions, such as renal agenesis or anencephaly, will not have a certain outcome. Part of our work with women and couples involves exploring how they feel about the uncertainties and unknowns and their individual capacity to deal with an outlook that is difficult to predict.

Most parents are anxious to know how exactly the anomaly will affect the quality of life for their child and what kind of impact it might have on themselves and their family. They may weigh up other factors, such as their support network, chosen lifestyle, and work and financial situation. Some religious faith and personal values will play a part in their decision-making, but many will find that beliefs they previously thought unshakeable will shift in the complex reality of prenatal diagnosis. This can add to the sense of loss parents feel, as along with the loss of the "healthy" baby they had envisaged, they also have to grapple with their changing sense of self.

ARC sees its role as ensuring that women and couples have the information and support they might need to make the best decision they can in their individual family circumstances. Help with information gathering may include signposting to relevant clinical expertise and details of specific support organizations (dependent on the condition diagnosed). ARC will also help them negotiate internet searches for reliable sources of information. Once they have gathered all the information possible, we will do our utmost to help them accept their final decision and reassure them that in our long experience parents are able to live with the consequences.

Although ARC does not offer professional counseling, the helpline team (currently three) use counseling techniques when interacting with women and couples by phone and face-to-face. Perhaps the most important skill we employ is effective and active listening to ensure parents feel listened to and know that their story matters to us. The team is trained to respond in a warm and measured way, allowing callers the time and space to explain their situation and try to work out what it means to them and the possible next steps. Reflecting back what has been heard is a useful way to help callers clarify their thinking, particularly when they are often in emotional distress after the shock of a diagnosis.

> *To this day I can still remember every phone call we had with you, how incredibly patient you were, incredibly helpful, understanding, nonjudgmental, relaxed and totally approachable. We were able to ask things we felt we could not ask the hospital staff. We were able to ask the questions that frightened us and those to which we did not want to hear the answers. We never met any of your staff face to face but had some of the most difficult and meaningful conversations we have ever had, over what remains the most difficult and precious time of our lives.*
>
> ***E-mail to ARC, October 2016.***

For couples who are struggling to come together on the decision about the future of the pregnancy, we can work with them on speaker phone or at a face-to-face session if they can travel to our London offices. For women or couples who find themselves unable to work out how to proceed or have a more extreme psychological reaction to their circumstances, we can signpost them to psychological services who can provide specialist intervention.

Karin Diderich

As HCPs, we should be aware that what we as HCPs say can have a large impact on couples, good or bad. I heard of a couple who went to a psychologist who was not experienced in dealing with terminations of pregnancy who asked the couple "what such an alien baby looks like." You can imagine this was far from helpful for the couple. What is helpful for couples is to hear about other couples' experiences, to know that they are not the only ones going through such a difficult period.

Once I went to visit a couple shortly after the termination of pregnancy to do a physical examination on their child and found them in the company of the grandmothers. I complimented them on their truly beautiful son (who had a lethal skeletal dysplasia) and performed the physical examination. A while later I learned that this couple had expressed their gratitude for my reaction. I spoke from my heart and did not consciously realize how much these parents would value that.

Frederike Dekkers

Main recommendations

- Allow for time to let information of possible anomalies sink in.
- Repeat information but avoid unnecessary medical terms.

- Be available and guide couples to reliable information they can read and share with others.
- One key HCP who knows the file completely.
- Provide practical information and prepare couples for what is coming: Which evaluation procedures are going to be carried out? What will couples hear during different appointments?
- Give words to the burden couples feel.
- Show humanity, empathy, and acknowledgment. This is of tremendous value.

References

Aite L, Zaccara A, Mirante N, et al: Antenatal diagnosis of congenital anomaly: a really traumatic experience? *J Perinatol* 31:760–763, 2011.

Best S, Wou K, Vora N, Van der Veyver IB, Wapner R, Chitty LS: Promises, pitfalls and practicalities of prenatal whole exome sequencing, *Prenat Diagn* 38:10–19, 2018.

de Wit MC, Srebniak MI, Govaerts LC, Van Opstal D, Galjaard RJ, Go AT: Additional value of prenatal genomic array testing in fetuses with isolated structural ultrasound abnormalities and a normal karyotype: a systematic review of the literature, *Ultrasound Obstet Gynecol* 43:139–146, 2014.

Dekkers FHW, Go A, Stapersma L, Eggink AJ, Utens E: Termination of pregnancy for fetal anomalies: Parents' preferences for psychosocial care, *Prenat Diagn* 39:575–587, 2019.

Evans C: *Genetic counselling: A psychological approach*, Cambridge, 2006, University Press.

Hammond J, Klapwijk JE, Hill M, et al: Parental experiences of uncertainty following an abnormal fetal anomaly scan: Insights using Han's taxonomy of uncertainty, *J Genet Couns* :2020.

Hui L, Bianchi DW: Noninvasive Prenatal DNA Testing: The Vanguard of Genomic Medicine, *Annu Rev Med* 68:459–472, 2017.

Jones K, Baird K, Fenwick J: Women's experiences of labour and birth when having a termination of pregnancy for fetal abnormality in the second trimester of pregnancy: A qualitative meta-synthesis, *Midwifery* 50:42–54, 2017.

Kaasen A, Helbig A, Malt UF, Naes T, Skari H, Haugen G: Acute maternal social dysfunction, health perception and psychological distress after ultrasonographic detection of a fetal structural anomaly, *BJOG* 117:1127–1138, 2010.

Lafarge C, Mitchell K, Fox P: Termination of pregnancy for fetal abnormality: a meta-ethnography of women's experiences, *Reprod Health Matters* 22:191–201, 2014.

Lewis C, Hill M, Silcock C, Daley R, Chitty LS: Non-invasive prenatal testing for trisomy 21: a cross-sectional survey of service users' views and likely uptake, *Bjog* 121:582–594, 2014.

Lou S, Lomborg K, Lewis C, Riedijk S, Petersen OB, Vogel I: "It's probably nothing, but..." Couples' experiences of pregnancy following an uncertain prenatal genetic result, *Acta Obstet Gynecol Scand* 99:791–801, 2020.

Minear MA, Alessi S, Allyse M, Michie M, Chandrasekharan S: Noninvasive Prenatal Genetic Testing: Current and Emerging Ethical, Legal, and Social Issues, *Annu Rev Genomics Hum Genet* 16:369–398, 2015.

Sandi C: *Stress and cognition* 4:245–261, 2013.

Skirton H, Patch C: Factors affecting the clinical use of non-invasive prenatal testing: a mixed methods systematic review, *Prenat Diagn* 33:532–541, 2013.

Sullivan HK, Bayefsky M, Wakim PG, et al: *Noninvasive prenatal whole genome sequencing: Pregnant women's views and preferences* 133:525–532, 2019.

CHAPTER 7

The iterative process of decision-making

Sam Riedijk[1], Frederike Dekkers[2], Lara Ras[3], Jane Fisher[4], Eva Pajkrt[5], Liesbeth van Leeuwen[5], Iris Emons[3] and Karin E.M. Diderich[6]
[1]Medical Psychology, Clinical Genetics Erasmus Medical Centre, Rotterdam, The Netherlands
[2]Psychotherapy, Private Practice, Rotterdam, The Netherlands
[3]Experience expert, The Netherlands
[4]Antenatal Results and Choices (ARC), London, United Kingdom
[5]Department of Obstetrics and Gynaecology, Amsterdam UMC, University of Amsterdam, Amsterdam, The Netherlands
[6]Department of Clinical Genetics, Erasmus Medical Centre, Rotterdam, The Netherlands

Karin E.M. Diderich

Diagnosis, counseling, and decision-making

This is the timeframe during which couples hear the (most likely) diagnosis and have to make the decision to terminate or continue the pregnancy. This timeframe is often very short. During the course of only a few weeks, couples see a large number of health care professionals (HCPs) and have to learn to interpret new and complicated information. With this information, they, as (future) parents, have to make a decision with widespread and extensive consequences. In the Netherlands, couples are helped by a multidisciplinary team that determines the recommended option for perinatal management in the case of pregnancy continuation. In Chapter 2, From the Consulting Room: Personal Narratives of Giving Prenatal Information About Fetal Anomalies, Wildschut, Bijma and Knapen set out how such a multidisciplinary team operates, and how it supports couples in their decision toward terminating or continuing a pregnancy.

After making the decision, couples need to be well informed about all the procedures that will follow. Therefore this timeframe is very extensive and time-consuming. It does not stop when the decision to continue or terminate the pregnancy is made. In this timeframe the couple also needs to be prepared for the termination of pregnancy procedure, allowing them to start making a scenario for how they want to handle this. HCPs should collaborate with each other to provide couples with all the necessary information and medical and psychological care.

Lara: Mae

We knew that a lot was wrong with our baby and the possibility that we were merely facing an orthopedic problem was soon ruled out. But we still did not know the exact nature of the problems our baby had. Did our child suffer from a syndrome? If so,

Prenatal Genetic Counseling
DOI: https://doi.org/10.1016/B978-0-12-823329-0.00004-0

what syndrome? What would be the implications for our child's life? Would our child even be viable? What would be the impact on our lives? What is intolerable suffering?

Almost immediately after our first meeting with the clinical geneticist, we got so thoroughly entangled in all our questions that we were offered a session with a medical psychologist. We spoke with her every week. In the meantime, I tried to keep up with all the information and insights that we gathered by registering everything in a notebook—each meeting with a doctor, all the research we did ourselves—the little black book. I needed this to somehow transform the tangle of theoretical information into a narrative that I could understand and grasp.

In the meantime the result of the rapid aneuploidy detection test came in. The results were normal and showed our baby was a little girl. We felt so much joy, we were having a girl! A boy would have made us just as happy, but somehow knowing the sex of your child makes it so much more real. We also felt joy because of the favorable test results. And we also felt sad because we still did not know whether we could and wanted to let the pregnancy continue. Hugo and I spent a lot of time talking with each other, our parents, and people from our social surroundings who we expected to have valuable insights to contribute to our decision-making process based on their life experiences.

I noticed that, to me, it became extremely important to receive some sort of test result as the foundation of my decision. Merely knowing that a lot was wrong was not enough for me. I wanted to know whether our daughter would have a real chance to live or whether she would pass away. I so desperately wished Mother Nature would decide for us and that she would pass away during pregnancy if she really was that affected. But unfortunately things were not made easy for us.

The clinical geneticist kept going through all the test results and ultrasound anomalies and kept us updated all the time. She seemed to understand my longing for a diagnosis and visibly put a lot of effort in finding it. Finally, she deducted a likely diagnosis: campomelic dysplasia (CD). To ascertain this diagnosis a DNA test needed to be done; however, such a test would require 6 weeks to generate a result, and we did not have that much time. Still, this seemed to be a plausible diagnosis. The clinical geneticist shared everything she could find about CD with me. Often this would be a publication from a scientific medical journal, which was almost impossible to understand for me. Still, she did her best to make me understand, by highlighting parts and by summarizing. Also, whenever she promised she would call me, she always kept her promise, even if that meant calling me in the evening, when she was at home. It is awful to be so extremely tense and anxious and then having to wait for a phone call about a test result or answer that comes later than promised. In this situation, we had so little time to make a decision that each day felt crucial. The dedication the clinical geneticist showed us was a huge source of support. I truly felt that we were in this together, that we were a team. I would highly recommend her approach to caring for us as well as communicating with us to other HCPs; it helps patients a great deal.

In the meantime, our worst nightmare became reality. The whole genome microarray analysis revealed no anomalies. I remember very well that in one of the conversations with the medical psychologist, I stated not knowing if I would be able to live with myself if I would decide to end the pregnancy based on a wrongful diagnosis. I desperately needed a definitive diagnosis before feeling able to decide whether or not to terminate the pregnancy. That final diagnosis, however, could not be established before 24-week gestation and we had to make a decision. After extensive deliberation, we decided, although not entirely convinced, to end the pregnancy. During this period of decision-making, we went to the movies to find some distraction. The movie we saw featured a character named Mae. It was some sort of Western movie, and we considered Mae to be a real cowboy name, a name that would fit a little girl with crooked legs that made her look like a little cowboy. That settled it, her name was Mae.

Frederike Dekkers

Psychological challenge for couples during this timeframe

> *How do we decide to continue or terminate this intensely desired pregnancy and live with our decision for the rest of our lives?*

In some cases the diagnosis and prognosis are quite clear. The child is not viable or the anomalies found are so severe that living without excessive handicaps, limitations, or pain is simply impossible.

The majority of couples with this kind of diagnosis and prognosis decide to terminate the pregnancy—unless they cannot reconcile this with their personal beliefs or moral standards (e.g., religion, culture, or ethics). Couples are devastated, but knowing that they can prevent their future child from leading a life of great pain and disability typically eases the couples' minds. Or couples mention that they do not wish to burden siblings with the care of a severely handicapped brother or sister. Even though couples face a very painful and difficult period, they reflect on this timeframe as a time of gaining clarity about the anomalies found and of being able to make a decision because of the information they obtained. Couples appreciate a clear summary of all fetal congenital abnormality and genetic conditions that are found and conclusions regarding the severity and prognoses being drawn by a multidisciplinary team. The fact is that all specialists involved communicate with each other and are aware of the steps taken that are elements in which the couples take comfort. Couples are also grateful if the number of specialists involved in their care is limited. A small number of specialists help to keep communication clear and limit the chance of the couples receiving contradictory information.

Knowing what to do, for the sake of their unborn child, their other children, and themselves, puts an end to their feelings of uncertainty. They can then "move on" to the next phase—that of preparing for the delivery.

When the diagnosis is hard to determine

There is also a scenario in which the precise diagnosis and prognosis are difficult to determine. Technology to detect fetal congenital abnormality and genetic conditions has developed greatly, but the result of all the diagnostics can still be insufficiently conclusive (Claustres et al., 2014). Phenotypic expression of the intellectual or physical disability of a child with a specific condition can range from mild to severe. Couples consult different specialists, search the Internet, and receive information from family and friends. It is difficult for them to interpret and evaluate all this knowledge. They crave for solid and comprehensible information. Will my child be able to walk? How much surgery will my child have to undergo? How much pain will my child have to endure? How much lung capacity will my child have? Will he/she be able to play sports with friends? While these are all valid questions, the specialists are not always able to provide an exact answer. In Chapter 2, From the Consulting Room: Personal Narratives of Giving Prenatal Information About Fetal Anomalies, Wildschut, Bijma and Knapen show why this is so complicated—see Box 5 "Issues hampering rational decision-making following the detection of major structural anomalies by ultrasound examination." The less conclusive diagnosis and prognosis are, the harder it is for specialists to answer some of the questions couples have. By contrast, the inconclusiveness of the specialists' findings reinforces the couples' need to hear those answers, as they are the ones facing the dilemma.

Some HCPs believe that the disability as such is not the primary issue. It is not a matter of the child being able to walk or not; the main concern is that the couples need to be fully aware that their image of a healthy child is gone and whatever the disabilities are, their lives are going to change drastically. I do not entirely agree. Couples have to base their decision on a future scenario, an image of themselves as parents and as a family. That future scenario must be one they envision themselves able to cope with. They understand that specialists cannot present a 100% true picture, but they have to base their decision on some kind of facts. That is why, in my opinion, couples hang on to statistics or the answer to the question "will my child be able to walk?" Wildschut, Bijma and Knapen call this "scenario making" in their chapter. They invite couples to envision what the consequences and implications are of the different options, at the level of daily life for the child, themselves, and the family. I very much agree with this approach, and I think imagining different scenarios helps the couples—both during this timeframe and going forward.

Some couples reflect that they would even have liked to receive more data, more statistics, to help them decide (Lalor et al., 2008). This supports couples in feeling that the decision is not so much their own personal choice, but one based on "facts as much as possible." This craving for facts underlines how taxing it is for couples to have to make this decision.

Sam Riedijk

Decision-making

Grieving starts in this phase, when couples lose the carefree image they had of their child (Statham et al., 2000). When the advanced ultrasounds and additional diagnostic tests are set in motion, results start to come up that start feeding into the decision-making process of the couple. While grieving necessitates letting go, it also entails getting a hold on adjusted images of their child. Each ultrasound or test result may provide new information. Couples struggle to get a grip on quickly changing images of their child. *At first I thought it would be horrible if my child would be missing a hand. Now I can only hope that will be the only issue. I think my child can have a worthy life with one hand.*

Each image requires attributing meaning to. What does living with this anomaly mean to my child? To me? To our family? Do we feel this is a life worth living? Can we organize the lives of our other children, our jobs, around the needs of this child? Couples need to confer with themselves, discuss with each other, and with HCPs. Couples often report not getting a lot of sleep in this period of time. Feeling consumed and overwhelmed with information and quickly changing images and their feelings.

Decision-making should comprise an exhaustive list of considerations, arguments, emotions, and information. In my experience, when couples discover an argument or experience an emotion that has not fed into the configuration of the final decision, couples experience a lot of distress. Instantly, couples doubt the validity of their decision. Would we have decided to continue or end our pregnancy if we had known that...? If we had thought of...? This can be anything, even seemingly "light" considerations. Knowing that there is a medical day-care center around the corner where the child could have received adequate care. Experiencing the overwhelming love for the child once it is born, discovering a similar case that turned out fine, and also receiving the actual diagnosis months after the decision can be a very taxing experience, causing couples to repeat the entire decision-making process. *Had we known that this was the diagnosis, I fear that we might not have made the same decision.* Those couples who end up with concluding the decision would have been the same can find peace again. Those who do not, start to have to deal with intense regret, regardless of whether or not this could have been known during the decision-making process.

Knowing this makes it a topic to address during counseling (Hodgson and McClaren, 2018). HCPs can advise the couple to start to make a document that comprises all facts, feelings, arguments, and considerations that underpin their decision (see scenario writing). This also provides a tool to experience some grip, to create an overview of what couples already know, and also of what they need to learn more about. Considerations are often fed by imagination. Unraveling imagination from concrete information can provide insight into what information is still needed and what not. For instance, couples hardly ever search for information about medical daycare. In the Netherlands, such facilities exist. Imagining your child may go there is something different from actually visiting such a medical daycare center. Such activities are powerful tools to gather a "real feel" of life with a child that has a medical condition.

In the long term, such a document can also help couples remember how they have reached their decision and enable them to conclude again that they have done everything within their power to reach their best fitting decision.

I must admit that most of my encounters with couples were with couples deciding to end their pregnancy, which is caused by the fact that I work at the end of the line. The couples who decided to continue their pregnancy despite severe anomalies resembled in their argumentation, which is in line with scientific literature (Hurford et al., 2013). It is helpful to quote those couples, so that a deciding couple can feel what resonates in them. *The only life my child can live is in my belly, and it is not up to me to end this life any earlier.* Or *I have lost a husband to a congenital heart condition. But I would have not wanted to not have known him. Now that the child in my belly also turns out to have a congenital heart defect, I am sure that giving birth and spending a few precious moments with my child will be worth the life. I cannot live with myself if I had to end it now.* In these cases the anomalies in the child were most likely lethal, and couples were preparing for loss during pregnancy or soon after birth.

Quotes of couples ending a pregnancy are equally important in making this scenario more tangible. *I rather feel the pain of losing my child, than that my child needs to live a life of suffering.* Or *I am sure that we can take care of this child, but what happens when we are gone? I do not want to put this burden on our other child(ren).* An elegant way of offering a quote could be as the following: "In my experience, I have often heard couples state....[quote]...How do you conceive of that?" This approach can also be used to summarize a number of considerations that are important to feed into the decision-making process, such as how life can be organized practically if the child would need hospital treatment or would never learn to walk or how couples perceive the quality of life of their child, how the siblings will be affected by having to share attention and resources with the affected newborn, and how willing or able the couple is to adjust their current lives to the requirements of the affected child, for instance, in terms of career, spare time, or depending on the nature of the anomaly, also things like holidays. Sometimes couples feel embarrassed when they are weighing the importance they attach to family holidays to the existential right of their child. It helps when an HCP tells them that other couples take these aspects of their lives into consideration, and that each consideration is valid to reach the best possible outcome of the most important decision of your life.

An essential consideration is the couples' ability to bear the responsibility of deciding to end or continue the pregnancy. Some couples anticipate that seeing their child suffering is something they cannot cope with. Conversely, some couples feel that they could not bear the responsibility of ending a pregnancy. When one of the partners states that ending the pregnancy feels like murder, then this individual may never be able to forgive him/herself for this decision. In which case, it is not a sustainable decision. In my experience, I have not yet encountered a couple who has successfully overcome regret, which, in turn, is something I share with couples who are in their decision-making process.

The question that I always begin my counseling with is "Although I have read your medical file, I would like to hear from you in your own words, what you hope or expect to bring about in this session." Couples often respond that it is their first time with a psychologist and do not know what to expect. But they hope to acquire some strategies to gain grip, to deal with this decision-making process. Sometimes working with couples in this time of their lives feels like teaching couples to speak the language of decision-making, painting the landscape they are in, and handing them the words they can employ to start the iterative process of decision-making.

In this sense the needs of couples resemble the needs of people experiencing grief for the first time (Maguire et al., 2015; Robinson 2014; Cassaday, 2018; Cacciatore and Thieleman, 2019). When overwhelmed with feeling things that you have never experienced before, when it seems you are overtaken and lost, realizing that you are in an entirely different landscape, where what you feel is the rule rather than the exception can be highly reassuring. It provides a frame of reference that allows a person to assess what is normal, what to expect, what words can be used, and what may be the dot on the horizon.

When couples are not in-line

The legal position of the pregnant woman can be the cause of confusion, also to HCPs. In the end, the decision is legally of the pregnant woman to make, since it concerns her body. However, psychologically, most women do not perceive the decision as one of which she is the sole owner. HCPs may be inclined to address the pregnant woman more during counseling, and even partners quite often state that they will follow whatever the pregnant woman decides (Harding et al., 2020). Most pregnant women indicate that, although they understand the good intention, they mostly tend to feel deprived of support when this happens. The child is a cocreation of both woman and partner or is a result of a deliberated decision of both partners. The decision to either continue or end the pregnancy is generally perceived by pregnant women as a shared responsibility, a decision that they cannot "take over" from their partner. HCPs can validate the shared responsibility of the couple by addressing both partners during counseling and by not assuming that both partners will speak with one voice. For both partners to be able to reach a future proof decision about the pregnancy together, it is essential that both partners are engaged in the decision-making process.

Sometimes couples are not in-line. Oftentimes, but not always, a pregnant woman is set out to continue the pregnancy, whereas the partner is more inclined to end the pregnancy. In the book of decision-making, both partners are writing different chapters. The pregnant woman representing the stakes of the unborn child and her attachment to it, whereas the partner more often represents the stakes of the entire family, the other children, and the need to keep the world spinning. Stating that together, partners represent the whole picture can be a powerful intervention to validate both partners simultaneously. In the process of decision-making, both partners need psychological space to explore both possible scenarios. If one of the partners is predetermined for one of the two, this tends to block the space of the other to explore. Exploring then feels like conceding to the wish of the partner, instead of feeling that it could be one's own decision. Often I have witnessed during counseling that a partner was describing feeling scared of regretting ending a pregnancy, and that would be the invitation for the partner to start recounting numerous reasons why continuing the pregnancy is not an option. Both partners need psychological space for exploring both options, and both partners are capable of granting each other that space. In my experience, most couples are willing to, once they discover how. Couples may find themselves going through the same discussions over and over, trying to convince each other or trying to make the other understand that what the other wants is really impossible.

As an HCP, counseling a disputing couple can be challenging, and a number of pitfalls are easily tread. It helps to realize that the process of decision-making is the responsibility of the couple. As an HCP, we can try to offer support, but we can never take over. It is, therefore, useful to listen to couples in terms of process instead of content. This implies that in listening to a process, one can hear that each partner is fixed in their preferred option, instead of the one stating not to be able to see her child suffering, and the other stating that he could never forgive himself for ending a pregnancy. Content can invite an HCP to feel more understanding for one of the partners. Content can also invite the HCP to attribute more validation to one argument over the other. However, once an HCP starts to do this, the relationship becomes triangulated. One partner can feel left out, whereas the other can feel more empowered. This has no benefit, since the couple is not taking the HCP home with them. And once they are together again, the HCP has probably caused more distance between the partners than helped them to engage in the decision-making process together.

One of the interventions that has often helped couples in my experience was to agree together that neither of the partners would state their preference aloud anymore. Instead, both would start to work on the scenario document, to write down all facts, expectations, thoughts, emotions, and deliberations for each of the scenarios. Together, they would work on the document, seek out additional information, discover the areas that had not gotten enough attention yet, etc. Only when both partners would feel that their document was complete, they could look each other in the eye and speak out their minds. I have often heard back that this approach worked well, and that partners felt relieved to have a procedure to follow, in which all of both partners' input was necessary and valid. In a few cases, even this procedure could not align couples.

Sometimes there is no room for compromise in core values. In those instances the question what the highest price will be has sometimes worked. A partner once said to his wife: "I really feel that our family is under heavy threat here. I think that the two of us are unable to manage this situation. Please, let us end this pregnancy and try again for a healthy pregnancy. I know that you will need to grieve, and I will be there to support you." The trouble is that if a partner ends a pregnancy without being sure of this decision, grieving can be impossible. She then told him that if she would not be able to forgive herself, she foresaw that she would never be herself again.

In their exasperation, it sometimes happens that couples will ask the HCP: What would you advise us? Or even What would you do? Remaining nondirective here implies that an HCP will never state (s)he would continue or end the pregnancy. Giving such an answer would invalidate one of the two scenarios and would make the HCP coresponsible for deciding. No good can come of this. Couples will frequently go back to this decision process to repeatedly evaluate the validity of their decision. In those instances the HCP will not be present to add his or her share. Taken out of the configuration, the couple may come to the conclusion that perhaps they would not reach this decision again, if it was not for the statement of the HCP, and that would almost certainly open the gateway to regret. Instead of answering what you would do, as an HCP, it is much more beneficial to provide the couple with examples of decision-making deliberations and quotes of other couples. And also by expressing that there is no one best decision, only the decision that best fits this couple.

In the case of the couple that could not align due to their conflicting core values, time was running out. The procedure for ending the pregnancy was planned at 23 weeks and 6 days gestation, in case that would be their decision, which was on a Monday. The last counseling session was the Friday before. I asked the pregnant woman what she would answer if the gynecologist would ask her on Monday whether she was sure about ending the pregnancy. I explained to her that this happens, because no HCP will ever take on the responsibility of ending a pregnancy if a woman is not certain. She replied that she would say that she was not sure. She would actually end her pregnancy leaning on her partner's conviction. This was a painful realization for her partner. Since she had no intention of lying to her HCP, the only possible outcome was to continue the pregnancy. Upon this realization she began to cry, feeling guilty to impose this decision on her partner. In contrast, her partner felt relieved. He had done everything he could and there was clarity. Months later, I ran into this couple in our hospital. He was carrying the infant carrier and was a very proud father. In hindsight, both felt that they had reached the right decision.

There is not always a happy ending, and there is not just one road that leads to decision-making. What is always essential though is that all cards are put in the table, and that the HCP manages the process of decision-making and not the content. In this case the couple did not need to grieve after deciding to continue the pregnancy. Instead they needed to learn to adjust to living with a child that needed additional medical care.

Reaching a decision

Sometimes reaching a decision seems like an impossible mountain to climb, especially in countries where a time limit is posed on pregnancy ending. It helps couples if their counselor voices aloud that a large majority of couples succeed in reaching a sustainable decision together, and that at this point there is no reason to believe that this would not apply to them as well. Expressions of hope and faith are powerful interventions to promote autonomy.

Frederike Dekkers

Decision to continue the pregnancy

Some couples reflect that once the diagnosis was clear and severe anomalies were found, they did not feel much room for any other decision than to terminate the pregnancy. Some express that they felt pressured into deciding to terminate the pregnancy by their HCPs. For some couples, arguments, such as the health of the mother, that there is "no more use for continuing" the pregnancy, are very painful to hear. This is especially the case when the couples' religious or cultural beliefs conflict with a termination of pregnancy. Couples point out that they would have wanted more information about what would happen if they would decide to continue the pregnancy. What are the possibilities if the child is born, what will the (short) life of their child be like, how will it influence the practicalities of their own lives, and what kind of support and comfort care will they have access to? It is highly recommendable to discuss the scenario of continuing the pregnancy, even if it seems to be a clear-cut case to the HCPs. We need to avoid couples doubting themselves in hindsight for not thoroughly thinking out the possibility and scenario of continuing the pregnancy.

Jane Fisher

Continuing a pregnancy after a diagnosis

Antenatal results and choices (ARC) is the only UK organization that has specialized support services for parents ending a pregnancy after a diagnosis. For this reason, we have more ongoing contact with parents in this circumstance. It would be impossible for one organization to have the capacity to address all the postnatal and long-term needs of women continuing pregnancies with a wide range of diagnoses and potential outcomes. However, ARC regularly supports women through the pregnancy after diagnosis as the emotional support on offer can be valuable.

> *It seems like you go places, and feelings don't really come into anything. It's fact all the way through. Um, and feelings tend to take over in a situation like this. Your feelings are actually a lot stronger than the facts. The facts are there, and having been told by the surgeon they'll do everything they can when it's born, then that's a fact. But how to deal - it's really how to deal with the feelings that go with it (Statham et al., 2001).*

The safe, independent space provided by the ARC helpline means expectant parents can offload anxieties about what can sometimes be an uncertain or changing outlook. The helpline team has a good knowledge of which condition-specific support organizations in the United Kingdom might be able to give specialized information and peer support. ARC can also check that parents are being seen by appropriate clinicians and encourage them to ask the questions that are important to them. These questions can include whether there might be prenatal interventions or therapies available and the ARC helpline team can help them investigate this and help manage their expectations about their eligibility and the potential outcomes.

ARC produced a handbook for parents continuing pregnancies after diagnosis in 2003 (latest edition 2018). The focus of this booklet is not on the diagnosis or outlook for the baby but more on the needs of the expectant parents, as we had evidence that these were often overlooked in the inevitable interest in how the pregnancy and birth will be clinically managed in view of the findings. For expectant parents who know their baby is likely to die at or shortly after birth, we have trained peer support volunteers who have shared similar circumstances.

Methods of termination

Eva Pajkrt and Liesbeth van Leeuwen

Appointment to discuss induction of labor for termination of pregnancy: structured interview important issues to discuss from a fetal medicine specialist point of view

Before explaining the procedure of termination of pregnancy, it is important to discuss the options on how pregnancy can be terminated, so that shared decision-making can be done. Basically, there are two options. The first is a dilatation and evacuation (D&E), usually performed in a specialized (abortion) clinic. The second is an induction of labor with the use of drugs in a hospital setting. Some women will opt for a D&E either in or outside the hospital facilities, because they prefer a quick procedure under general anesthesia. It depends on the country of origin where the D&E will take place. In the Netherlands, most hospitals will offer the option of D&E up to a gestational age of 13 weeks. In abortion clinics, this is available up to 22 weeks, but only in the absence of severe medical conditions in the pregnant woman that would warrant a hospital delivery. Dutch gynecologists are not trained to do a late D&E or the so-called Finks procedure. In contrast, in the United Kingdom, several hospitals offer D&E. Chapter 1, A Flow From Screening to Diagnostics, in Section I, provides an overview of the organization of prenatal care in a variety of countries. Induction of labor is often chosen because it entails the opportunity to see and hold the baby.

Frederike Dekkers

It is important for couples to know what will happen if they decide to terminate the pregnancy, knowing that their child is not viable, will not be born alive, or die soon after birth. They need information about how the baby will be taken care of after birth, what kind of comfort care is possible. What will the baby look like? How much chance there is that the baby will be born alive and which signs or symptoms should they anticipate? How long will the baby live? How will the baby die? This should be followed by more practical information. If they want, couples can bring their own blanket to put their baby in and their own little cradle. Conserving the baby in water is also a possibility. Couples should be informed about the ways their child can be preserved from birth until burial or cremation. Couples can contact a funeral home to discuss the possibilities. In the Netherlands, there are photographers who work for a volunteer institution whom you can call to come and take photos. Do couples want to make a birth/remembering card? Additionally, who do they want to be present at the delivery?

This is an intense conversation to have with couples, but one of utmost importance. This information and conversation will greatly alleviate the psychological process of seeing and saying goodbye to their child. Knowing what the options are and, therefore, being able to choose the way that fits these specific couples with this specific child is of enormous value. Couples have one shot at this. Information is a way to get at least some sense of control and direction in a process that leaves couples almost no control (Asbring and Närvänen, 2004).

Jane Fisher

Ending a pregnancy after a prenatal diagnosis

Across three decades, ARC has supported thousands of women and couples through termination for fetal anomaly. The helpline team can draw on this experience to explain what has helped other parents through this difficult time. We can offer reassurance that our extensive anecdotal evidence, along with research literature, suggests that the majority of parents find a way to cope with and manage this painful life experience (Korenromp et al., 2005).

Once women and their partners have made the decision to end the pregnancy, ARC helps to make sure that they are as prepared as they can be for what might be ahead, practically and emotionally. This includes offering them a copy of our booklet for those facing termination. ARC first produced its "handbook" in 1990 (latest edition 2017) because women facing termination told us that their information needs were not always met.

The need for the handbook was clear when women described not being prepared for the length of time a medically induced labor could take or the level of pain involved. Many were shocked that they delivered a recognizable "baby" and did not feel ready for how it looked. Some were not helped to decide whether they wanted to see and hold their baby. Those women whose pregnancies were beyond 13-week gestation were often shocked to learn that their hospital could only provide medical induction as the method of termination (due to a lack of in-house expertise in performing surgical abortion).

The ARC handbook gives women and their partners the opportunity to gain an understanding (in their own time and at their own pace) of what the process involves so they are better equipped to discuss their preferences with their health-care team. It can help them work out the method of termination (medical or surgical) that is most congruent with their coping style. It gives them a sense of what to expect during and after the procedure and some of the options they have. This can give them some control back at a time when the psychological shock of the diagnosis can make them feel they have little. The booklet can be downloaded as a PDF from the ARC website, provided in hard copy by National Health Service (NHS) units, or sent free of charge by post from the ARC office.

Iris: Noud

The next day another gynecologist from hospital number 3 performed the amniocentesis. I remember requesting to see the ultrasound image to check whether everything was really that bad. I felt this need because we had been told so much in the past days. Maybe they were wrong, maybe they had not seen it right? She replied by stating that the sonographer who had done yesterday's ultrasound was a highly experienced and competent professional, and that if she had seen these things then they must be there. After the puncture she said Nobody would blame you if you were to decide to end this pregnancy. We see a multitude of anomalies at this point and the chances are that your child will have severe learning disabilities. That is often the case when a child has a syndrome.

The amniocentesis did not generate any explanation for the anomalies, but we decided to end the pregnancy based on the poor prospects for our baby. It had to stop, I was drained. To cherish hope every day only to feel it bashed away each time. How could anything good come out of this pregnancy? We both were exhausted and we wanted it to end. There was no room available in hospital 3 and I had to wait 5 days because there was yet another national holiday and a weekend. The gynecologist understood that this was intolerable to us and arranged an earlier date in hospital number 1 for us. This meant that I could be admitted within 2 days.

I made the decision, we made it together and we felt with all our heart and ratio that it was the right thing to do. No doubts, but a lot, a tremendous amount, of sadness.

Eva Pajkrt and Liesbeth van Leeuwen

Reflection period

Since we are dealing with life-defining decisions of pregnancy termination in generally very desired pregnancies, the care provider should allocate enough time to discuss the abovementioned options. Moreover, in some countries, such as in the Netherlands, once the decision to have a termination of pregnancy is expressed, it is mandatory to have a (5-day) reflection or consultation period, before the appointment can be scheduled. This reflection period begins the day of the woman's referral to a hospital or abortion clinic by the general practitioner or the day of self-referral by the woman. In the case of fetal anomalies the onset of the diagnostic workup is generally regarded as the start of the reflection period. Consequently, once the couple definitively express their wish to terminate a pregnancy after workup is completed, the mandatory 5-day reflection period has usually passed. However, if a prenatal diagnosis is clear-cut and does not need additional workup, for instance, in the case of anencephaly, the reflection period remains mandatory. The World Health Organization (WHO) is of the opinion that waiting periods may undermine women's rights and autonomous decision-making and stresses that countries should not impose medically unnecessary delays in abortion care (Europe's Abortion Laws: A Comparative Overview, 2020). This may be true in socially indicated abortions where reflection time may restrict timely access and affordable care. In our experience, however, a mandatory reflection period in the case of a fetal anomaly is generally well accepted by couples and usually beneficial to the grieving process. When a fetal anomaly is detected, couples are often overwhelmed, and their first impression might be to end the pregnancy as soon as possible. But terminating a pregnancy within a day of diagnosis will not make things easier or make the pregnancy disappear. Grief is always associated with a termination of pregnancy and taking time to reflect on the decision and discussing it together and with friends and family will eventually prove favorable after birth. If the sequence of events from diagnosis to termination is fast, the couple may be left with the feeling of a roller-coaster ride that completely took over and allowed no time for emotional adjustment. In general, it is difficult for men and women to reach 100% certainty about the termination of a desired pregnancy, as it is too emotional. The task of the HCP is to protect them from regret in the long term. Therefore we like couples to support their decision for at least 97%–99%. In addition, women often relate the anomalies found to something they did or did not do. It is, therefore, important to actively inform a woman that it is not her fault.

When setting the date for the termination, it should be discussed with a couple that a waiting period allows reflecting, and the positive side of this period should be stressed. Of course, this should never be exaggerated in the case of severe maternal psychological distress.

What could have helped Iris:

- Nobody inquired whether I would like to talk about this with a social worker or psychologist. All of the conversations were rather clinical. I needed to talk with someone to address the emotional aspects of our decision, a safe space where we could work on finding meaning to all the clinical results and the decision we had to make.
- The gynecologist arranged the termination of pregnancy within 2 days. However, in hindsight this was too fast. Someone should have told me that it is okay to wait for 5 days. That you need 5 days to prepare and give shape to ending the pregnancy. Someone experienced in the matter should have given us advice regarding how to prepare emotionally for the pregnancy ending and the period after.

References

Asbring P, Närvänen AL: Patient power and control: a study of women with uncertain illness trajectories, *Qual Health Res* 14:226–240, 2004.

Cacciatore J, Thieleman K: Normal Complications and Abnormal Assumptions After Perinatal Death, *MCN Am J Matern Child Nurs* 44:6–12, 2019.

Cassaday TM: Impact of Pregnancy Loss on Psychological Functioning and Grief Outcomes, *Obstet Gynecol Clin North Am* 45:525–533, 2018.

Claustres M, Kožich V, Dequeker E, et al: Recommendations for reporting results of diagnostic genetic testing (biochemical, cytogenetic and molecular genetic), *Eur J Hum Genet* 22:160–170, 2014.

Europe's Abortion Laws: A Comparative Overview. 2020. Accessed 09122020, at https://reproductive-rights.org/europes-abortion-laws-comparative-overview.

Harding E, Hammond J, Chitty LS, Hill M, Lewis C: Couples experiences of receiving uncertain results following prenatal microarray or exome sequencing: A mixed-methods systematic review, *Prenat Diagn* 40:1028–1039, 2020.

Hurford E, Hawkins A, Hudgins L, Taylor J: The decision to continue a pregnancy affected by Down syndrome: timing of decision and satisfaction with receiving a prenatal diagnosis, *J Genet Couns* 22:587–593, 2013.

Korenromp M, Christiaens GCML, van der Bout J, et al: Long-term psychological consequences of pregnancy termination for fetal abnormality: A cross-sectional study, *Prenatal Diagnosis* 25:253–260, 2005.

Lalor JG, Begley CM, Galavan E: A grounded theory study of information preference and coping styles following antenatal diagnosis of foetal abnormality, *J Adv Nurs* 64:185–194, 2008.

Maguire M, Light A, Kuppermann M, Dalton VK, Steinauer JE, Kerns JL: Grief after second-trimester termination for fetal anomaly: a qualitative study, *Contraception* 91:234–239, 2015.

Robinson GE: Pregnancy loss, *Best Pract Res Clin Obstet Gynaecol* 28:169–178, 2014.

Statham H, Solomou W, Chitty LJ: Prenatal diagnosis of fetal abnormality: psychological effects on women in low-risk pregnancies, *Best Practice & Research Clinical Obstetrics & Gynaecology* 14:731–747, 2000.

Statham H, Solomou W, Green J: *When a baby has an abnormality: A study of parents experiences*, Cambridge, 2001, University of Cambridge.

CHAPTER 8

Ending the pregnancy

Eva Pajkrt[1], Elisabeth van Leeuwen[1], Lara Ras[2], Frederike Dekkers[3], Jane Fisher[4], Iris Emons[2] and Sam Riedijk[5]
[1]Department of Obstetrics and Gynaecology, Amsterdam UMC, University of Amsterdam, Amsterdam, The Netherlands
[2]Experience Expert, The Netherlands
[3]Psychotherapy, Private Practice, Rotterdam, The Netherlands
[4]Antenatal Results and Choices (ARC), London, United Kingdom
[5]Department of Clinical Genetics, Erasmus Medical Centre, Rotterdam, The Netherlands

Eva Pajkrt and Liesbeth van Leeuwen

Dilatation and evacuation

Worldwide, together with aspiration, dilatation and evacuation (D&E) is still the most frequently used method of termination of pregnancy. There is some evidence that such terminations have an association with cervical insufficiency and premature birth and this should be discussed with the couple (Lemmers et al., 2016). Alternatively, the administration of vaginal misoprostol a couple of hours before the D&E is also being advocated, specifically for this reason. A potential downside of this procedure is that the couple will not be able to see and hold their baby and, thus, have the option to say goodbye. On the other hand, there is some evidence that does not show negative effects of this procedure (Hennegan et al., 2018; Ryninks et al., 2014). In addition, some couples do not wish for any ceremony after birth and that should be respected as well as both favorable and unfavorable effects of seeing and holding the baby have been reported.

Induction of labor

Mifepristone 36 hours before hospital admittance

Termination of pregnancy is usually performed with drugs, which is nowadays a combination of mifepristone and misoprostol. There is strong evidence that the administration of mifepristone subsequently followed by misoprostol at least 1 but preferably 2 days later is an effective and safe method. Initially more toxic drugs were used for pregnancy termination, such as hypertonic saline or sodium, that had to be injected intraamniotically. This was associated with an increased risk of intrauterine infection

Prenatal Genetic Counseling
DOI: https://doi.org/10.1016/B978-0-12-823329-0.00008-8

(Ho, 2017). Alternatively, labor was induced with high levels of sulproston intravenously. Not only was this synthetic prostaglandin E2 derivative contraindicated in a large number of women, but its use was also associated with many side effects, leading to much discomfort during an already anxious situation. The introduction of mifepristone, an antiprogestin blocking the action of progesterone at the receptor level and misoprostol, also a synthetic prostaglandin E2 derivate, registered as a stomach protector, to prevent gastric ulcers, and not as abortion tablets, positively changed the practice of termination of pregnancy. Nowadays, misoprostol is used all over the world for this purpose. Misoprostol can be administered vaginally, orally, or rectally (but not intravenously) and is prescribed in various treatment regimens with varying doses. Prior to the induction of labor with misoprostol, the use of mifepristone will sensitize the myometrium to the action of prostaglandins, thus facilitating the onset of labor and significantly contribute to a shorter delivery. Once hospital admittance is scheduled, the patient is administered one tablet of mifepristone approximately 36 hours before the scheduled induction of labor. Not much is to be expected from this first tablet, but in the case of fetal demise, mifepristone may trigger labor. Other side effects are headache, skin rash, stomachaches, nausea, or blood loss. If possible, provide these pills in hospital. Tell people in advance that taking the mifepristone pill is an emotional moment since it is the beginning of the termination process and advise them that they should preferably not be alone. If the woman gets contractions or blood loss after taking mifepristone, she should contact the hospital.

Hospital admittance

On the day of the scheduled induction of labor, the couple is admitted to the hospital. Consider admittance time with the onset of the day shift, to avoid an unnecessary number of care providers. On admittance the first dose of misoprostol is given vaginally. It is very important to tell the couple in advance that the duration of the delivery is unpredictable and will last an average of 12 hours, with 90% of women delivering within 24 hours, but that it sometimes takes more than 1 day. Make sure that the patient delivers in a room where she is away from other women in labor, although you cannot avoid that couples see other pregnant women. In most cases a single or delivery room is used. The latter may be confronting but allows for all facilities that are also used in a premature delivery to be at hand. The possibility of having an accompanying partner, family member, or friend to stay over in the room as well should always be offered.

When informing women about the induction in the case of termination, it should be stressed that the pattern of labor is different from what may be known from previous deliveries. Instead of contractions that come and go, there may be a continuous pain and pressure that reaches its maximum just before the baby is delivered. Moreover, effacement and dilatation of the cervix is also a less gradual process than when giving birth to a living term baby. The cervix may be closed till the very end and suddenly just burst open. It is often not necessary to reach 10 cm (full dilatation), as the baby is usually small and can already pass before full dilatation is reached. Therefore it is not necessary to evaluate dilatation every 2 hours as in term labor. If pain becomes unbearable and pain relief is requested, it is good practice that the care provider takes time to sit next to the patient to evaluate the situation, assess whether delivery is imminent, and evaluate effacement, prior to administering pain relief. Once the cervix is fully dilated, pushing out the baby is usually not so much of a problem. However, in the case of anxiety, progress can be difficult and gentle encouragement should be used. In most cases, it is not necessary to expedite the delivery and sufficient time should be given to the women to give birth according to their own needs. The care provider should be aware that when regular contractions or severe pain suddenly disappear, the baby may be already expelled in the vagina and the woman needs to push to deliver. Women are often advised to urinate in a bedpan instead of going to the toilet when labor has really started, because of fear of sudden delivery.

Assessment of the presence of fetal heart before the termination

Sometimes, the anomalies of a fetus are so severe that intrauterine demise may well be expected within a short time frame. In those cases, we advise to check the presence of the fetal heart rate before the D&E or induction of labor. The awareness that "nature has taken its course," and that an anomaly was indeed lethal, will usually help the grieving process. In the aftermath of a termination of pregnancy, the fact that a fetus has died by their own decision is usually one of the most difficult aspects for couples to deal with, sometimes leading to lifelong guilt. The knowledge that the fetus died without their decision can be invaluable.

Side effects and pain

The most common side effects from misoprostol are nausea, stomachaches, and fever. Pain is different from pain during a term delivery. Pain is very difficult to predict, as women will undergo labor pain knowing that they will not have a healthy baby as reward. Therefore the pain experienced can be worse than what women remember from a previous delivery. Below 20 weeks, epidural anesthesia is usually not necessary and morphine injections or similar painkillers are preferred. Termination of pregnancy is more a sensible decision while emotionally the woman may feel that the baby is safe with her, not ready to let go. Morphine sometimes helps literally to make the influence of the mind less strong and facilitates the process of letting go of the baby and thereby speeding up the delivery process. On the other hand, from a psychological perspective, women have indicated that the delivery happened in a blur due to the morphine. Also, morphine is not always effective in pain alleviation (Andersson et al.,

2014). Not having been "present" psychologically during the delivery is something that may interfere with grieving afterward, as it makes it more difficult to realize that this event really happened.

Lara: Mae

We tried hard to prepare for the delivery. We had conversations about how the delivery would be induced and we were told that most of the times the delivery took around 12 hours. We discussed about how we would take our baby home with us after the delivery and we decided that we would bring her home in a straw basket. I looked for some clothes in supersmall size but I decided to make a cover blanket on which I embroidered some trees, an elephant, and a giraffe.

I was given a pill to swallow that evening. Even though Hugo and I really made this decision together, swallowing the pill made me somehow feel more responsible for the decision than Hugo. It was my body that held our daughter, and if I did not take the pill or simply chose not to be admitted to the hospital then nothing would happen. To date, this is a burden that pushes hard on me.

I was admitted to the hospital on a Saturday. I was at 23 weeks and 6 days gestation. The gynecologist seemed to be stressed about that and let us know the baby should be born before 24 weeks by law. (The gynecologist was interpreting the law incorrectly. Labor needs to be induced before 24 weeks gestation.) We were given a single room in the obstetrics ward. Hugo had a stretcher in this same room so that he could stay with me throughout the delivery. It was so incredibly painful for me that my room was adjacent to rooms filled with pregnant women who were admitted for pregnancy complications. They received visitors, their husbands, and their children. There also were a few mothers and newborns. The radiator in our room was broken and the temperature was so high that we had no choice but to leave the door open. It was hurtful to be confronted with all of these women and couples who would be leaving the hospital with a living and healthy baby, whereas our baby would be dead. I do hope hospitals will one day arrange for separate wards for women who have to go through this, so that they will be able to experience their pain without being confronted with what they lost.

Labor was induced and I had to lay down for a while. I do not know why but I got scared of getting back on my feet. In the meantime, I experienced no contractions whatsoever and, therefore, the dosage was increased. The only effect was that I became nauseous and had to vomit a lot.

When nothing changed after 24 hours, the medication was changed and I was given morphine. It gave me a cramped belly but no contractions.

After the next 24 hours had passed, an ultrasound was made to check how the baby was doing. The sonographer asked me whether I wished to see the baby and whether I wanted to know if she was alive or deceased. I did not want to see her and I did not want to know whether she was alive either. Until this day, I regret this decision, and until this day I wonder whether she was alive at that moment or not. I was panicking during this ultrasound. We were told that most deliveries took around 12 hours, why was mine taking so long? How come the delivery did not start? Was my body trying to tell me that I had made the wrong decision? The gynecologist said that this sometimes happens and that this delivery takes a long time to start, and if I really did not want to proceed with ending the pregnancy, we could opt to go home. At that moment I was terrified that the baby was already deceased and that I would go into delivery at home and be all by myself. We stayed....

After the ultrasound a balloon catheter was placed. I was sobbing intensely when this was inserted. I remember so clearly that the woman doing it asked me why I was crying. The tone she used was stern instead of empathic and that hurt me deeply. The preceding days I never cried, and, therefore, this question felt more like an accusation, making me feel misunderstood and alone.

Hugo had transformed into a caged tiger by then. He did not want to leave me alone, but he had to take air and move around after all these days in the room. I just wanted to stay in bed and do not remember what I did to pass time.

In the night from Monday, on Tuesday, I started to feel some pain in my belly and on Tuesday morning my water broke. The dilatation was sufficient for this tiny little baby to be born. This last part of the delivery went smoothly and there all of a sudden, on February 14, 2012, we became the proud parents of Mae.

I had mixed feelings after giving birth. I was proud and felt a lot of joy about having become a mother, but I was also anxious and hesitant about seeing her. I did not want her on my bare chest; therefore I kept her on my t-shirt. Although I knew she would not be a beaming pink baby at this term I was a little bit shocked about the way she looked. The nurse in service was a man called Ben and he assisted us during the delivery.

He took our pictures, and after birth he inspected her together with us. He told us that he could see that she must have died a little while ago. He pointed out her "spiderfingers" and odd-looking genitals. To me it felt very good to closely inspect her this way, together with Ben. Ben had seen many babies, also premature babies and he spoke with us without any taboo or hesitation about what he perceived. Now that Mae was born, she showed so many more anomalies than those we had seen on the ultrasound. We spent some time with Mae in her cover blanket on my lap and we made many more pictures.

We felt a little uncomfortable about Mae being deceased for a while already and, therefore, we decided to hand her to the nurse and Mae was put in the cold store. The nurse took her hand and foot prints and wrote her date and time of birth and her birth weight on the cards.

Unfortunately, my placenta had not come out 1 hour after the delivery so I had to go into surgery. It was decided that I would have an epidural but I refused this. I was exhausted and sad and I did not want to have to listen to what the surgeons said to each other during the surgery. I wanted a total anesthesia and they granted me this. I was relieved.

I lost a lot of blood during surgery. Therefore I had to stay in the hospital longer to be monitored. That Tuesday afternoon my mother and Hugo's parents visited us at the hospital to meet Mae. We experienced so much pride and sadness at the same time.

I was given a blood transfusion and then we could go home, at last. Mae stayed behind at the hospital for autopsy on Thursday.

When we arrived home, Hugo and I went straight to bed, we were broken. I remember I cried so intensely. The blood loss had made me very weak and I needed time to recover physically.

We invited our friends and family to come to our house on Friday and celebrate the birth of Mae, and to say farewell. On Friday morning, we returned to the hospital for Mae. The hospital staff prepared her wonderfully, wrapped in the cover blanket I made for her, and they had even added a little teddy bear. Hugo and I put her in her straw basket together and we took her home.

At home we had beautifully decorated a little table, on which we put the hand and foot print cards, a candle, and beautiful flowers. We put the basket on the table and made some more pictures. Then we closed the basket and our visitors arrived. We ate biscuits with pink decoration, as is tradition in the Netherlands, and we toasted to Mae with pink champagne.

I think at that moment we celebrated our pride and joy foremost over Mae's arrival. After a while Hugo and I went to the crematorium with our close family, where we spent some time in the family room together, and then we left Mae behind. She was cremated that same day.

The first days after Mae's birth we kept a little book in which we wrote down our conversations with Mae, and the questions that we wanted to ask Mae. To us, that came closest to having a real conversation with her.

Hugo went back to work after a few days, which ensued that I was home alone. I could not work or go out because I was very low in energy. During that period I was quite active on a support website called "Little angels." This was a support group for parents having lost a child. I came into contact with other mothers who were in similar situations. The first months after Mae's birth I had a hard time sleeping. Especially during those nights I went online and wrote long testimonials of my sadness in this support group.

Frederike Dekkers

Preparing for the delivery and birth

If couples decide to terminate the pregnancy, they need solid information about how the delivery will be induced. If parents decide to induce the labor by drugs (as described above by Eva Pajkrt and Liesbeth van Leeuwen), women needs to take the medication at home, before the admittance to the hospital. This is to help facilitate the delivery. Women can feel guilty about taking this medication, even going as far as saying that they have killed their own child. It is very necessary to explain in detail what the reason is for this medication and how it will help women deliver their child. What should be emphasized is that it has nothing to do with them killing their unborn child. Lisy et al. (Peters et al., 2016; Lisy et al., 2016). found that empathy, warm and sensitive communication from the health-care professionals (HCPs), coming to a shared decision, respect from us as HCPs for the individual needs and preferences from couples, and provision of understandable information about diagnosis, birth, and the period after birth are reported as being very helpful for couples.

What can we offer couples?

Couples look back on this period and are very grateful for proper, solid, and good information about the fetal congenital abnormality or genetic condition found. They appreciate understanding and compassion from the HCPs they encounter. HCPs that show humanity convey the message that they see the couples as human beings rather than (just) another case. This can be achieved by showing interest into how the couples feel, and how they are coping so far—besides explaining the medical information. Ask what they need to be able to make the decision: "How can we help you make this difficult decision?" Truly sympathize with their enormous dilemma and task as (future) couples to make this, most feared, decision. It also helps couples tremendously if HCPs know their file, their context, not just now, but possibly also any past experiences. For some couples, this is not the first loss. This requires a lot of time and effort from us as HCPs. We need to be aware of the difference we can make for couples in the years after the birth of their lost child. It matters enormously how we intercommunicate with couples (Lisy et al., 2016).

Couples are very grateful for the time they are given by HCPs, to be able to ask all their questions, even when they have already asked them (twice). They also show their gratitude for HCPs offering their availability. An invitation to call or email them whenever couples need to is of invaluable worth. Also, couples appreciate specific knowledge and honesty about the abnormalities, or the condition found. They know they are the ones who have to make the decision, but they feel very supported if the HCP confirms their decision or line of argumentation. They even say this helped them in the months after the termination. Some couples reflect that they appreciate a little guidance from the HCPs in the decisions, others appreciate the neutrality and honesty of the specialist. It is up to us as HCPs to ask, and sense, what kind of guidance couples need from us and what suits us as human beings and professionals to be able to give couples.

Psychosocial care

Couples reflect that being asked if they needed more psychosocial care helped them to evaluate their needs in this specific period. It helped them to know that support was available if they needed it. Couples also reflect that they regret rejecting psychosocial care that was offered them. Ramdeney et al. found that only half of a group of women were aware of available psychosocial care resources and had thought about their individual need for support (Ramdaney et al., 2015). Half of the couples in their study did not anticipate any need for psychosocial care and rejected psychosocial care both during and after a termination of pregnancy. Couples run from appointment to appointment and are trying to get their heads around everything they hear. Feeling what kind of psychosocial care you need is, therefore, hard. It is advisable to point out to couples that it is difficult for them to know what their needs for psychosocial care will be, but it is always available. This should continue to be a topic of conversation within the appointments, and it is highly advisable that HCPs keep repeating the offer for psychosocial care.

Couples report that in most cases the medical care they were given was thorough and of a high professional standard. But they missed a conversation about "how do we make a decision like this as parents," reflecting with them about the process of making a decision of this kind. This is where psychosocial care should collaborate with medical care, in order to provide couples, the whole range of care in this great life-changing dilemma that they are facing (Dekkers et al., 2019).

Contact with other couples

Some couples would have liked to have been in contact with other couples in a similar position or who have gone through a similar process. Worldwide, numerous websites exist where couples can exchange experiences and ask each other questions. Some organizations set up meetings for couples and grandparents and organize memorial services.

Future pregnancies

The moment couples realize their pregnancy is not going to result in the child they so much wanted, they start asking themselves when they can start getting pregnant again. Couples feel guilty about these thoughts, but almost all couples have these thoughts at some point during this process. Even though the pregnancy did not end well, their wish for a child remains. What I have learned from couples is that the wish for a child may even become more persistent. Couples express a desire to discuss a future pregnancy (Dekkers et al., 2019). Which risks they will encounter, how couples can prevent this from happening again, how will they be supported in a future pregnancy, and so on.

Case N

Mother N is a 28-year-old woman who, after delivering a healthy daughter, had decided with the father not to have a second child. But 2 years later, the mother turned out to be pregnant while using the contraceptive pill. The mother instantly decided that she wanted to keep the baby. The father wanted to do a noninvasive prenatal test (NIPT), expressing that he did not want a handicapped child. The mother reluctantly agreed to the test but did not share the father's opinion, saying a child with a handicap is also welcome. The couple decided to do a basic NIPT, not an extensive one, because their midwife told them they could get into trouble with taking out a mortgage if certain anomalies are found in the NIPT. (This is incorrect information.) At the time, they were in the process of obtaining a mortgage, so they opted for the basic test. No abnormalities were found in the NIPT. The mother was feeling uncertain about the pregnancy—she had not bought anything for the new baby boy—but the test results calmed her down.

At the 20-week ultrasound, an anomaly was seen in the heart of the baby and the couple was referred to a tertiary medical center with the instruction "not to worry." At the extensive ultrasound and further diagnostic evaluation, a Tetralogy of Fallot in combination with a chromosome aberration (6q15q21 deletion) was diagnosed. At first, the mother still thought it would be ok. But after hearing the HCP saying that although her baby had a heartbeat, he was not "actually" alive, and that he would not be able to function on his own, she first began to think about terminating the pregnancy. After a few days, the couples decided, in harmony, to terminate the pregnancy.

The mother was referred to me because she decided, after having much doubt, to witness her baby going into the cremation oven. She cannot let go of that image and has not worked since her son was born. Furthermore, the due date has not yet passed and, in her mind, the mother feels as if she is still pregnant. She cannot bear to see her sister because she is pregnant, this hurts too much.

What can we learn?

The couple did not quite agree on the arrival of a second child and the terms under which a child was welcome in their family. This made the first weeks of the pregnancy very lonely for the mother. After further diagnostic evaluation, it helped this mother to hear that their child was affected greatly and would probably never function on his own; he would always need help in his daily life. It was because of this explanation that she could decide to terminate the pregnancy. She did not want this for her son. She needed the HCP to set out the scenario for her.

Deciding to witness the cremation turned out to be the wrong decision for this mother. Of course, the effect that certain decisions have cannot always be predicted. HCPs need to point out to couples that treatment can help to overcome a troubling image. Some couples simply do not know that such treatment, like EMDR, exists. We need to prevent couples from carrying these burdens and enable this mother to continue working and take care of her daughter in the way she wants.

We should keep in mind when the due date will pass. Especially mothers tend to mentally continue their pregnancy of which the due date remains a significant date. Furthermore, we must be aware how much hurt it causes some couples to see others being pregnant or to see young babies.

Hospitalization and delivery in the case of induced labor

When the decision is made to terminate the pregnancy, the stage of preparing for hospitalization and delivery has arrived. Couples arrive at the hospital with contradictory feelings. They are facing what they most dread: delivering a stronly desired child and knowing that at the same time they are going to have to say goodbye. It is an event they fear very much, but with the proper help and guidance from us as HCPs, it can become an event couples can look back at with a sense of peace.

One mother wrote to me: "It may sound a bit crazy, but my husband and I look back at the day of the delivery with warm feelings. Everybody was so kind and they prepared us so well. We both felt it was right." This is a compliment to all the HCPs involved with these couples.

Psychological challenge for couples in this time frame

How do we meet and say goodbye to our baby at the same moment?

A solid preparation in the previous stage (emotional and practical) is of tremendous value when ending the pregnancy. But still, couples have to come to the hospital and start the delivery that they fiercely wish was not there. They are going to have to let go of this life they so much wanted in their lives.

Offering couples all the time they need when they arrive at the hospital is mentioned as one of the most valued factors in this stage. It helps couples grasp the reality of the moment, but even more, time helps them to process this parting in their own way. HCPs who realize what an immense task couples are facing and who can really voice this understanding offer invaluable support. Preparing for the upcoming hours together, actively asking couples how they want this process to go and reassuring them if they feel insecure, all help to guide the couples through this phase.

Being able to guide couples in this way requires training. Perinatal HCPs should be aware of their central role in helping couples and should seek and be provided with appropriate training (Ravaldi et al. 2018). At the same time, investing in couples not only professionally but also emotionally makes HCPs vulnerable to compassion fatigue: emotional and physical exhaustion. During this time frame, this is particularly true for the nurses who spend most time at the couples' side (Hutti et al., 2016). It is advisable to let compassion fatigue be not only a recurring topic in intervision with colleagues, but also in performance reviews. HCPs, too, need support and to *feel* supported.

Personal feelings of the health-care professional

An HCP has a right to his/her own opinion about the reason for termination, whether this is based on professional experiences or personal beliefs. But if that opinion contradicts the couples' decision to terminate the pregnancy, this HCP should be well-aware that having a strong personal opinion on this topic will likely influence the way that they communicate with couples, and how they are perceived as care givers (Richardson and Ormond, 2018). Nonverbal communication especially is almost impossible to eliminate. HCPs in that situation should ask themselves the question whether they should be the ones taking care of these couples. Doing your work professionally is not enough, one *must* also be involved emotionally (Limbo and Kobler, 2010; Werner-Lin et al. 2016; Engler-Gross et al., 2020).

Jane Fisher

Work with health care professionals

Since the mid-1980s, ARC has accumulated a wealth of feedback from women and families regarding their experience of prenatal diagnosis and the accompanying quality of care. It remains a priority to share this intelligence with HCPs to explore what might constitute best practice. We do this through ARC's professional training program.

ARC runs three well-established study days attended by doctors, midwives, sonographers, and genetic counselors. Day 1 covers communication skills in the context of antenatal screening and diagnosis with an emphasis on delivering unexpected or difficult news; day 2 explores how best to support decision-making after prenatal diagnosis and the importance of coordinated and individualized care pathways; and day 3 uses a number of case studies to examine how to achieve high quality care in more complicated scenarios. All 3 days comprise a mix of evidence based and experiential learning with maximum interaction. There is acknowledgment through all our training that this is a challenging area of healthcare both professionally and personally and we emphasize the importance of staff support and self-care. Health professionals are invited to make use of ARC's confidential helpline.

ARC runs training days from its London offices and is regularly commissioned to run regional days and "in-house" tailored workshops for staff and lectures for students.

In recent years the uniqueness and value of ARC's services have been recognized internationally. This has led to ARC staff leading training workshops in the Republic of Ireland, Belgium, Sweden, and Japan.

Frederike Dekkers

What can we offer couples?

When couples look back on this phase and are asked what was of most value for them, kindness, understanding, and empathy from the HCPs around them are most often mentioned. Sensitivity, empathy and attunement to each individual parent are also reported as valuable in a study by O'Connell et al. Couples place a great deal of importance to their interactions with HCPs and often have a very specific recollection of those interactions (O'Connell et al. 2016).

Couples report that they appreciate having one primary caregiver, a nurse, who is experienced in attending to couples about to give birth because of a termination after prenatal diagnostics. A caregiver who is not uncomfortable with the situation at hand and knows how difficult and special this delivery is going to be. Someone who is within reach and really makes contact with them.

Couples very much appreciate if the involvement of other caregivers or other workers in the hospital (cleaners, suppliers, etc.) is kept at a minimum. This primary caretaker can help couples take the cape or blanket they brought and remember them what they intended to do after birth. They can signal the moment when it is time to make lasting memories such as hand and footprints, nametag, and if couples want to, call the photographer.

Couples mention they found comfort in receiving an extensive explanation of everything they could expect to happen within the next 24 hours. From the hours leading up to the delivery to what to expect after delivery.

Seeing and holding the baby

There have been many studies about whether seeing and holding a stillborn is healing in the process of grieving and developing psychological symptoms. The conclusion Ryninks et al. drew, and this coincides with my own experience, is that seeing and holding their stillborn baby provides couples with time to process what has happened (Ryninks et al., 2014). They not only have time to make lasting memories, which show to be of tremendous value (Jones et al., 2017, Ryninks et al., 2014; Johnson and Langford, 2015; Crawley et al., 2013), but also, time to look at their child, hold and caress him or her, and say goodbye. If they want, they can share this experience with family members. Ryninks et al. also found that the majority of couples are satisfied with their decision to see and hold their baby. But there was also fear of seeing the damaged or dead body. Couples should, therefore, be well informed and it should be a thought-out decision to see and hold the baby (Ryninks et al., 2014). A systematic review by Kingdon et al. also shows positive outcomes for couples who saw or held their baby, who were given enough time with their baby, and who were able to make enough lasting memories (Kingdon et al., 2015). If people fear looking at the baby an HCP can also offer to look at the baby first and sometimes cover parts that may be more difficult to look at and then look together if desired. It is advisable to ask whether couples want to name their child, and when the baby is born, to say the name out loud together. Naming and saying the name out loud represents the recognition and acknowledgment of this specific child. As not all parents benefit psychologically from seeing and holding their baby not seeing and holding the baby should always be part of the possibilities. Some parents in our clinic have recited that they did not dare hold or inspect their baby but much appreciated that the nurse held and took care for their baby instead of them.

Differences in mothers and fathers

Our attention in this time frame often goes to the mothers, being the ones who have to accomplish the delivery and receive treatment. Jones et al. found in their study that fathers were often overlooked by HCPs (Jones et al., 2019). During hospitalization, men feel neglected since all the attention goes to the mother. HCPs should be aware and attentive to the pain of fathers who face the same loss and include them as much as possible in the standard care Cacciatore et al., 2013. How will couples cope with this together? Acknowledgment of the father and his role, and talking about it upfront, makes overlooking the father less likely to happen. Lizcano Pabón et al. also studied the perspective of fathers after perinatal loss (Lizcano Pabón et al., 2019). They found that, for men, it is difficult to support their partner during their emotions and grieving process. SANDS has written material: "Mainly for fathers" which may be offered to fathers.

Need for practical information and more practical matters

The need for practical information of other couples' experiences seems almost inexhaustible. Couples indicated that they have not only missed information about contractions when it is the first delivery, but also because it is an induced delivery. The possibility of getting sick and vomiting because of the induced labor made some mothers feel very disturbed, and this made it harder to focus on the delivery. In addition, information about how much blood loss to expect, upcoming milk production for breastfeeding, and what to do about that increased their anxiety. What options are there for pain control during the delivery? How much bed rest is needed after giving birth, even if the gestational age was "only" 17–23 weeks?

Giving couples all the time they need is of great importance. But the hours before delivery are also precarious. If couples have to wait too long before the inducing of the delivery starts, they mention becoming agitated by all the tension they feel about being in the hospital and the imminent delivery.

Couples appreciate grandparents being allowed to come visit, day and night, and that the father is allowed to spend the night with the mother in the hospital. It is also important to discuss what will happen to the baby when couples leave the hospital. Will they take the baby home or will they leave the baby behind in the morgue? Couples need to be prepared that the color of the skin of the baby will change if he/she has been cooled. The baby will look different after a couple of hours and even more after hours of cooling, which is necessary for preservation.

Couples prefer a room not surrounded by couples experiencing a "normal" delivery. They find it extremely confronting to see not only all the birth cards, photos of happy couples with happy babies, but also the sounds of a baby crying, heart monitoring, etc.

Couples very much appreciate it if their gynecologist or clinical geneticist visits them after birth. Couples feel seen, they feel important, and they see it as a sign of recognition of the existence of their child.

Couples also indicate that some humor and lightness helped them through these tough hours. Making it a little more bearable by using humor could release some tension. Being able to use humor in an effective way of course depends on the confidence of the HCPs. But one should not evade the use of humor if there is a clear opportunity.

In some cases, there is no member of the psychosocial care team available who is familiar to the couples—such as a psychologist or medical social worker that couples can speak to before, during, and after the termination. Couples expressed having missed speaking to someone they know and who knows their whole story and can guide them or intervene when necessary[16].

After delivery

After delivery, couples go home, with or without their deceased baby. In the Netherlands, couples can choose to arrange for the child's burial or cremation themselves or let the hospital take care of a cremation. There are some funeral services that specialize in the burial or cremation of (preterm) babies. They can help couples with all the choices they have to make.

A burial or cremation ceremony helps with the acknowledgment of a deceased child and, therefore, in coping with grief (Basile and Thorsteinsson, 2015). Couples reported more grief if they did not participate in a ritual for their deceased baby (Tseng et al., 2017). Such a ceremony also helps couples to share their experience with family and friends and to help open a conversation about their loss in the future. They have a shared experience to look back on.

Jane Fisher

Many parents describe loss after prenatal diagnosis as a very isolating experience. It is even less spoken about than other perinatal losses such as miscarriage or stillbirth, and fear of judgment can prevent some women and couples disclosing the nature of their loss to others. ARC has services in place with a view to breaking this isolation. For example, ARC has a network of trained volunteers across the United Kingdom who can provide peer bereavement support via telephone or email. These are women and men who have experienced termination for fetal anomaly for a variety of indications and at different gestations and subsequently offered to help others. To join this network, volunteers must be at least a year on from the termination and undergo a rigorous training program. The training is designed to verify that they have sufficiently integrated their own experience and have the necessary skills and capacity to support others. ARC's National Support Coordinator supervises the network.

Grief after termination for fetal anomaly can be complicated. Women describe feeling implicated in their baby's death because they consented to termination (Sandelowski and Barroso, 2005). Some will be struggling with conflicting feelings, including guilt (Kamranpour et al., 2020). Some believe they are undeserving of sympathy or support because they caused their baby's demise. The word "termination" for fetal anomaly is used (rather than the commonly used term "abortion"). This indicates a need to differentiate between the circumstances of ending a wanted pregnancy (that only becomes "unwanted" after the diagnosis) and "having an abortion" which is most often used to mean a woman simply does not want to be pregnant. The stigma surrounding abortion also adds to the sense of isolation some women and partners describe. They consider their situation different to those who have experienced stillbirth or miscarriage because, as they see it, the loss did not happen *to* them but was made to happen *by* them. To facilitate the sharing of some of these difficult feelings, ARC provides a password-protected online forum for women and men bereaved after a termination following a prenatal diagnosis. It offers a safe and welcoming space to share experiences, feelings, and coping strategies.

> *I could seriously have not got through terminating a sick baby without this website, helpline and specifically the forum - it was so good to know others were in the same situation and had coped.*
>
> *Your forum was such a support during the darkest time of my life. It provided so much reassurance to know I wasn't alone and it made me face what was going on and help me process it all.*
>
> ***(ARC service user audit 2018).***

Other services parents have told us they have found helpful are our newsletter and regional support meetings for bereaved parents. Three times a year we distribute approximately 2000 copies of ARC News in paper and digital editions. It contains women and men's accounts of their experiences, feature articles, and updates on ARC's work. We try to ensure that the parent stories reflect different types of experience and coping strategies, so readers might find their emotions normalized and validated. There is also a section for birth announcements of successful subsequent pregnancies.

> *I still gain great comfort from other people's experiences in the newsletter similar to my own which reduces the isolation especially as time passes and so few remember what I have been through.*
>
> ***(ARC service user audit 2018).***

ARC's regional support days enable bereaved women and men to meet face to face. The meetings are gently facilitated by an ARC staff member, and the program for the day moves from sharing of experiences and feelings to coping strategies and hopes for the future. As well as the support days we have a coordinator in Scotland who facilitates quarterly evening meetings in Glasgow. Comments in evaluation forms highlight the emotional intensity of the sessions but the opportunity to be with others who have a unique understanding and empathy is highly valued.

Example of ARC support: Case G

G contacted us by phone in late 2018. It was her first pregnancy and she had accessed cell-free DNA screening at a private clinic. She was shocked and confused when told over the phone that her test had thrown up an unexpected result. She was told the lab report said 47,XXY and was urged to contact her doctor; the person giving the result offered no other information. G contacted the laboratory where the cell-free DNA testing had been done and was fortunate to speak to someone who was able to tell her more about Klinefelter's syndrome and explain that it was important she had an amniocentesis. She went back to the hospital and had an amniocentesis that confirmed 47,XXY.

G contacted ARC having spoken to a geneticist who she felt did not acknowledge her concerns about what it might mean for her son and stressed the "mild" nature of Klinefelter's syndrome. G is a scientist and she and her partner undertook extensive research into the condition, reading studies, consulting the Klinefelter's society website, and speaking to a parent with a son with the condition. On the ARC helpline, she was able to be open and honest about what the diagnosis meant to her and her partner. She spent a number of calls and emails talking through what they had learned from their research and using ARC as an independent sounding board through the harrowing process of coming to a decision on how they wanted to proceed.

G and her partner finally decided that they needed to end the pregnancy. She was then enabled through conversations with ARC and by reading our handbook to recognize that surgical termination was the method she could best cope with. ARC helped her arrange this at an independent provider clinic. ARC's bereavement support services, such as our online forum and newsletter, were of comfort to her and her partner in the aftermath. ARC's support extended into her subsequent pregnancy and in February 2020, she gave birth to a healthy baby boy.

Eva Pajkrt and Liesbeth van Leeuwen

Birth

The couple should be told that above 20 weeks there is a small chance that the baby will be born alive. When this happens, it means that a faint heartbeat may be heard; however, the baby will not be crying or moving a lot. It should be explained to the couple that the process of dying postdelivery resembles an extinguishing flame and is not like going through a death throe. Although the thought of giving birth to a child that may be alive after birth is sometimes distressing, couples never mention a live-born baby as being traumatic afterward. As the baby is very immature, the aspect of the baby is different from what is known from a term baby: the skin is darker, sometimes a bit reddish, thinner, and the underlying veins are easy to see.

Depending on the law, it is advised in some countries to perform feticide when the gestational age is advanced above 22 or 24 weeks. Feticide ensures that the baby is not born alive. In our daily practice, feticide is only offered in the case of nonlethal anomalies beyond 24 weeks gestation.

Feticide is a procedure in which the fetal heart is stopped by an injection of potassium chloride intracardially or intrathoracically or in the umbilical vein. The procedure is performed transabdominally with or without local anesthetics by using a 20-G or a 22-G needle. Alternatively, in larger fetus, lidocaine may also be injected. In monochorionic diamniotic or monochorionic monoamniotic twins, selective feticide may be performed by cord occlusion of the affected twin.

Prior to the induction of labor, the couple should always be asked whether or not they wish to see the baby. It is our strong advice to do so. Couples tend to make the scariest images of their fetus in their mind and are usually relieved once they actually do see the baby. In the case of severe macroscopic anomalies, the confirmation of the prenatal findings by viewing the baby can sometimes strengthen the decision for termination and may also facilitate the grieving process. In contrast, the absence of external anomalies, such as a cardiac defect, may be confusing and caretakers should inform the couple about this. In the case of hesitance in the parents, the care provider can offer to view the baby first and subsequently view together with the parents. Sometimes the partner may be the first to look or vice versa. If a woman or her partner really does not want to view the fetus, we have to respect their wishes, since people have different needs and we are not all the same. However, we do advise to always take photos of the baby especially in those cases, which should be kept in the hospital files. It is not uncommon that couples call us, sometimes even years later, with the question whether we still have some pictures of the baby.

We always advise couples to bring a camera to make as many photos as they want, but most hospitals also make photos from the baby for the parents to have (Ramirez et al., 2019; LeDuff LD, 3rd, et al., 2017; Thornton et al., 2019). Some couples hire professional photographers to capture the few moments with the baby. If desired by the couple, hand and footprints can also be provided. To preserve the baby the "water method" is nowadays offered in many hospitals (Tiemens-van Putten, 2015). In the water the baby takes a natural position, the skin gets a lighter color, no malformation of the corpse occurs, and as a result it is easier to show the baby to other people or to make photos.

Box. Still society.

The Dutch Charity called Still was actually started by a pregnant woman who learned at 33 weeks gestation that her baby would not be viable after birth. She called birth photographer Melanie Plaggenmarsch for a photo shoot during pregnancy, and birth, in the hope that the baby would be alive long enough. The baby died after 1 hour and Melanie continued taking pictures.

The next day already, Melanie was contacted by the hospital for a similar photo shoot, and from that moment onward the requests continued to grow. She founded Still that currently works with 140 photographers and serves the families of 45 dying or deceased babies each week.

Families never need to pay for the photos. The photographers are paid by crowd funding and donations.

Still stimulates parents to welcome their child. "It is understandable that parents focus their attention to saying farewell. But parents become fathers and mothers in that moment. We encourage parents to make those moments beautiful, also by arranging for nice clothes, cover blankets, or basket."

The photographer makes memories, by taking pictures of little hands, feet, and eye lashes for instance; memories last for a lifetime.

Photographers may feel burdened by the events they record, as they are always intense. They need to support each other, and, therefore, they discuss each delivery with a coworker. A few times each year it happens that a photographer stops doing this work, but most photographers tend to continue. Photographers are not only the witnesses of sadness, but they are also part of the moment that couples become parents. They witness the pride and love for the child. "Imagine, a mother holds a lifetime of love in her heart but only has these short moments to share it with her child. I once heard someone say that nowhere is love so abundantly present as in a space where people say their last farewell. And I can only agree."

After delivery, couples are always asked if they wish to give their baby a name. Naming a stillborn baby acknowledges the existence of the child for many parents and helps them in their grieving process. A card with the name of the baby and the date and time of delivery is always provided.

After the delivery of the baby, the placenta is not born spontaneously in about 20%–30% of cases (Bryant et al. 2011). When assessing the placenta after birth, care should be taken to ascertain that the placenta is complete. Unrecognized incomplete placenta delivery may hamper the postnatal period, especially when women present with severe blood loss several weeks after the delivery or when blood loss continues well beyond 6 weeks after the termination. When the postnatal period is not without complications, the grieving process may be compromised, and women will often not be able to move on psychologically as long as the physical complaints persist. Not only in the case of a retained placenta manual removal is designated, but also in the case of doubt about completeness of the placenta, manual assessment of any retained placenta pieces is strongly advocated. When blood loss is under control, a maximum waiting period of 2 hours can be endured before removing the placenta manually in theater, often under general anesthesia. Since this procedure will separate the mother from her partner and baby, the waiting period allows time for spontaneous birth of the placenta and for the woman and her baby to bond. To a great deal the necessity of a surgical removal of the placenta will determine how long a couple may have to stay in hospital.

In general, couples may leave when everything is over, but some may prefer leaving the next morning during daylight after a night's rest. Some couples may experience a threshold to leave the hospital, as they are reluctant to leave the safe hospital "bubble." Going into the normal world again is a very confronting moment as life goes on for other people while the couples' world has turned completely upside down. If there are other children, the option of seeing the baby should be mentioned, as well as close loved ones. Especially when pregnancy duration is advanced, it is difficult for family or friends to imagine the impact of this loss, as they only know terms of miscarriage or term birth. The couple should also be advised in the period after birth to give each other some space as both partners go through their own grieving process.

Iris: Noud

The day we were admitted to hospital number one to give birth, I stood next to a pregnant woman. We both had a suitcase to carry what we needed for after the delivery; in contrast with me, she was also holding an infant carrier. She looked at me so lovingly, as if to say that we were in this together. Only my baby would be dead.

I was admitted to the ward where all women came to give birth to their healthy babies, albeit to a private room. You see the balloons, you hear the cheerful people and the crying babies.

The nurse asked us a lot of questions, like would you like to hear your baby's heartbeat one more time on the ultrasound, do you know if it is a boy or a girl, and do you want to give your baby a name?

It was overwhelming. And we answered NO to each question. We wanted nothing, just to get this over with and leave it behind us. We wanted this nightmare to end. We did not want to give our baby a name. We made that decision because we feared that otherwise hearing that name would always remind us of him. And that would only make everything needlessly complicated, right? Maybe we should not make this thing too big and just think what a lot of people have told us: from the start this was not a good pregnancy and this baby is simply not healthy.

We waited all day and I received a lot of pain medication because I had incredibly painful contractions. I had not anticipated experiencing the same kind of contractions I had when giving birth to Mats and it really hurt. I felt so many things but I did not want to feel anything. The dosage of pain medication was increased.

During the night at around 1:45, I needed to pee and the nurse got me a sort of wheel chair. They helped me on it and while I was sitting there I felt the baby coming. I started panicking, I was freaking out, I felt so scared, I started shouting "I don't want this. . .help me, help me I don't want this, this is terrifying!." In the meantime, the doctor arrived in my room and delivered the baby. It was so scary, I was not prepared. I went back to my bed with my partner and then the nurse sat on the bed holding the baby in her arms. We sat there for a while, looking at him.

He was small, red, and whole, with little nails on his little toes. We also saw that many things were amiss. The nurse asked us if we would like to touch him, but I was too scared, I did not have the courage. My partner touched him; he told me he was already feeling cold. The nurse wrapped him in a blanket and put him in a little bed. I never saw him since. But I could leave it behind me and I felt relieved.

The next day we returned home and tried to get on with our lives. But the relief that I felt at the hospital was short-lived. We fled, went from one vacation to another, anything not to be at home. I wanted to be together with my two guys, the big one and the little one. We were so happy to have the little one. Because of him we could carry on. I quickly resumed work.

What helps according to Iris:

- Explain beforehand why it is important to give your child a name, make pictures, and to hold your child in your arms. This is so important afterward.
- If people refuse, do not give up too easily, because it is highly important. Instead of perceiving their no as rejection, understand that couples are just not capable of grasping the importance of your suggestion at that moment. They have spent weeks in a rollercoaster, and they are exhausted and empty. Give them guidance.

What does not help according to Iris:

- I wish I had held my baby in my arms, I wish I had touched him. I know the nurse offered it to me, and I said no, but in hindsight I would have so desperately wanted to have done things differently.

- I wish I would have given him a name immediately and that would have really helped me to grieve.

Sam Riedijk

When couples reach the decision to end their pregnancy after careful deliberation, they most often experience some peace of mind. A new phase is entered, and couples start preparing for the procedure of ending the pregnancy. This procedure can vary strongly between countries (, 2020). In some countries, pregnancy ending is only allowed until 14 weeks gestation and is then performed with a surgical abortion. In others, pregnancy ending is allowed until 24 weeks gestation and induced labor is the procedure (Herziene NVOG-richtlijn, 2015). In other countries, Belgium and France, for instance, pregnancy ending is allowed throughout the pregnancy. Considering the viability of the fetus after 24 weeks gestation, the fetus will receive an injection intra-uterine to end the life before birth and then labor is induced. Since I am working in the Netherlands the experiences that I recount are in the setting of ending a pregnancy by induced labor before 24 weeks gestation.

When couples come to me to prepare for the ending of the pregnancy, I ask them what image they have of this event. Often this is mostly a medical image, a procedure. I will tell them that other couples in their situation have often shared their experiences with me, and that if they would like to, I could share these experiences with them. With few exceptions, couples would like to hear more about these experiences. I always recite the "average" experience of pregnancy ending. Almost always, this is the first time that couples start to feel what pregnancy ending means to them, and that it means that they are not only ending their pregnancy, but also that they will also become proud parents in addition to losing their child.

It is an extremely important intervention; to facilitate the transition from the medical experience of prenatal diagnosis to the parental experience of losing a child. If couples engage in pregnancy ending while still holding unto an image of ending a pregnancy, the delivery will come as a, likely traumatizing, shock.

Case S

In my first experience of offering psychological counseling in prenatal genetics, I encountered a woman who was pregnant with her first child. She was highly ambitious, worked at an accountant's firm, and she had a very rational and cognitive approach to decision-making. She discovered at the 20-weeks gestation that her child had severe congenital anomalies. After amniocentesis a trisomy 18 was detected, which meant that most likely her child would not survive the pregnancy or first period after birth. She and her partner were informed that they were entitled to end their pregnancy. The practical consideration that the child was not viable, and the emphasis on preventing further suffering resonated with her and her partner and they decided to end their pregnancy as soon as possible. She was afraid of what her child would look like, and her wish not to see her child after birth was honored. From their religious background, she was not allowed to attend the funeral, as this was seen as the task of the father and other men in the family. She was fine with that, as she wished to continue her life and hopefully fall pregnant again.

She did. No anomalies were discovered on the 20-weeks scan and she gave birth to a healthy girl. She came back to me shortly after, with attachment problems. She told me that giving birth to her healthy daughter had overwhelmed her with love and with motherhood. Having transitioned into motherhood, she could no longer understand why she had decided to end her pregnancy, and worse, why she had not chosen to see and hold her child. She was experiencing such intense, existential regret, that she could not allow herself to attach to her daughter. She felt she did not deserve her, that she was a bad mother. She described holding her daughter and thinking, if I had not ended my first pregnancy, you would not be here. What kind of person am I? What kind of mother would kill her child and not even see it? Her pain was not limited to psychological pain; she also experienced intense physical pains. Her love for her daughter made her feel love for her first child. From that love, she could not imagine ever wanting to end the life of her child sooner than the trisomy 18 would have caused.

What can we learn?

Preparing couples to the experience of the love they will feel for their child after birth is essential, because this is an emotional aspect of the scenario of ending the pregnancy, and that, therefore, should be part of the decision-making process.

When couples indicate that they would like to hear this "average" experience of ending the pregnancy, the following is what I describe:

What I am about to tell you is sort of the average of all the experiences couples like you have shared with me. Of course this means that it is in no way a prediction of what exactly it will be like. However, it does provide you with an insight into what pregnancy ending is, in the experience of other people.

When couples decide to end the pregnancy, they receive an appointment and a pill that the woman has to take in 2 days before checking into the hospital. Most women find it extremely difficult to swallow a pill that sets off the pregnancy-ending procedure. They tend to feel that this pill already stops the life of the baby in their belly. However, this pill does not end the life of the baby. It does weaken the cervix, so that induced labor is facilitated. But if a couple would decide to change their minds and continue the pregnancy, this would be possible after swallowing this pill.

After 2 days the couple is admitted to the hospital, in the same ward where women give birth to live-born babies, but in their own private room. They will be together and decide who can or cannot visit them. Labor is induced by medication. From all of the experiences that couples have previously shared with me, I can tell that giving birth for pregnancy ending is the same experience as giving birth to a live-born baby. The process can take as many hours, the contractions are equally painful.

Most couples, before giving birth, indicate that they are afraid of seeing their child, dead. However, most couples have told me afterward that they felt unimaginable pride and love upon seeing their child for the first time. In fact, just as overwhelming and intense as it is in the case of a live-born child. Even if the anomalies are visible on the outside and you would notice them, you will still feel that your child is beautiful and you will find family resemblances. The medical staff will respectfully care for your child, and if you consent, your baby will be put in your arms to hold and look at.

At this point, most couples are in tears. If that happens, I fall silent and ask the couple what they are experiencing.

Often the woman then replies that this is what she imagined, and it saddens her to lose her child, whereas often the man states that he had not imagined pregnancy ending to be like this. In his imagination, ending the pregnancy was something factual, a sort of black box. Both partners often state that listening to this "average" experience is important to them. Especially because they would not have wanted to be confronted with this experience only at the moment that it is happening. Sometimes this recital of the average experience turned out to be an intervention, where the couples realized that they did not foresee themselves being able to carry this out.

Case C

A couple presented at my clinic to prepare for ending the pregnancy. However, it turned out that the partner wanted his pregnant wife to prepare for the pregnancy ending, but she did not. Although she shared the decision to end the pregnancy with her husband, she did not share his desire to prepare for this psychologically. She was from Poland. As a psychologist, I knew that pregnancy ending is seen as a taboo in Poland and I imagined her refusal to prepare for this event was due to that taboo. Fortunately, I had attended a training by Fadua (see Chapter 5: Addressing Diversity in Prenatal Genetic Counseling *on diversity) and she had taught me to be very careful by discarding your professional authority in the face of cultural differences. She had given me the feedback in a simulated session that I was much too careful and that I should take the professional opinion and experience that I had very seriously, and try to find a way to relate to the person in front of me as best as I could. The woman in front of me was closed, resistant, agitated. I addressed her and said to her: "You will probably hate me for this, as I can tell that you have only one wish and that is to leave this room as soon as you can." She confirmed that indeed that was her state of mind. I explained to her why it was my conviction that she consider seeing and holding her child after birth, before rejecting it. I told her about the average experience, stressing the love she would feel for her child, and how important it would be for her afterward to have acknowledged her motherhood for her child. She was angry and anxious to leave, and of course I let her, apologizing for making her feel this way.*

It was a few weeks later that she came back to me, alone. She said: I deeply hated you when you insisted on telling me how much I would love my child, and how important it would be for me to hold and name my child. But you were right, and now, you are the only person in the world that I want to talk to. You are the only one who understands me. Thank you for having insisted. In hindsight, she blamed herself for not having taken her child home with her when she left the hospital. She returned to the hospital for her child the day after. She managed to work through that self-blame though.

What can we learn?

What I learned from the diversity training was to not discard my professional knowledge in the face of cultural diversity. I have often experienced thinking that my professional knowledge may not hold up or apply to other cultures. It made me powerless because this professional knowledge is all I have to offer. In this case, I allowed myself to practice with behavior that was recommended to me in a training but felt out of my comfort zone. Insisting on telling things the person in front of me was not ready to hear. This case has made me reconsider that it was worthwhile overcoming my discomfort for the long-term benefit of the person in front of me, all the while understanding that it is a fine line to thread. Even in writing this book I have discovered that for some people seeing and holding the baby does more damage than good. The fine line to thread is to take my professional experience and always leave room for the individual in front of me.

Eva Pajkrt and Liesbeth van Leeuwen

After birth

In the case of an uncertain or unknown diagnosis, a postmortem examination should always be advised. The manner of counseling and the urgency felt by the couple may have a great impact on the willingness of a couple to have this investigation. Most couples, however, find the thought that their baby will be "cut up" very stressful (Lewis et al., 2019). In the case of the suspicion of a skeletal dysplasia, a postnatal fetal skeletal status using radiography or computed tomography may be performed. When couples refuse autopsy, minimally invasive autopsy by postmortem MRI may sometimes be an alternative (Lewis et al., 2019). With or without autopsy a couple can arrange their own funeral of cremation according to their own wishes and beliefs. According to the law, burial or cremation is not mandatory before a certain gestational age. In the Netherlands the fetus can be left for collective cremation in most hospitals. Since a couple of years some hospitals also provide a donation service for the body for scientific research, but only if postmortem examinations are not necessary from a medical point of view. The knowledge that at least something "good" has come from a "bad" situation may provide couples with a sense of meaning and satisfaction. Prior to delivery, the couple should be informed about all these possibilities, and information on birth registration should also be given. If already desired, appointments for additional support should be scheduled.

Women are advised not to swim and bath (shower not included) or have intercourse as long as they have blood loss. They should get instructions on a normal pattern of blood loss after birth, which includes a decreasing amount of blood every day for a maximum of 3–4 weeks and to come when blood loss seems excessive or if they have a fever. Breast engorgement may occur, and tablets can be prescribed to prevent this from happening. Alternatively, a tight bra should be advised. A routine revision appointment is made for approximately 6 weeks after birth.

Frederike Dekkers

Main recommendations

Provide solid and comprehensible information.

Give equal attention to the option of continuing and terminating the pregnancy.

Show understanding, empathy, and compassion; couples need to live with this decision the rest of their lives.

When possible, offer couples a choice regarding the method of pregnancy termination.

Take time to know the file and the couples in front of you, so that you can focus on their reactions.

Be available for couples.

Confirm the decision couples make.

Discuss their need for extra psychosocial care.

Provide practical information about the upcoming delivery in the case of a termination and about (comfort) care at a neonatal unit in the case of continuing the pregnancy.

Discuss that couples need to think about their own way of seeing and saying goodbye to their baby after delivery.

And discuss the importance of a burial or cremation ceremony so couples can start preparing.

Provide couples with HCPs who are experienced in caring for couples who are going to deliver a child after prenatal diagnosis.

Allow couples time to say goodbye in their own way.

Prepare with couples, help them carry out the plans they made to meet and say goodbye to their child.

Be supportive, warm, and empathic, stay by their sides.

Make lasting memories with and for the couples.

Do not overlook the father while giving treatment to the mother.

Ask couples if they have a name for their child.

Give couples lots of practical information about the first hours, days, and weeks after release from the hospital.

Explain the importance of a farewell ceremony.

References

Andersson I-M, Christensson K, Gemzell-Danielsson K: Experiences, feelings and thoughts of women undergoing second trimester medical termination of pregnancy, *PloS one* 9:e115957-e, 2014.

Basile ML, Thorsteinsson EB: Parents' evaluation of support in Australian hospitals following stillbirth, *PeerJ* 3:e1049, 2015.

Bryant AG, Grimes DA, Garrett JM, Stuart GS: Second-trimester abortion for fetal anomalies or fetal death: labor induction compared with dilation and evacuation, *Obstet Gynecol* 117:788–792, 2011.

Cacciatore J, Erlandsson K, Rådestad I: Fatherhood and suffering: a qualitative exploration of Swedish men's experiences of care after the death of a baby, *Int J Nurs Stud* 50:664–670, 2013.

Crawley R, Lomax S, Ayers S: Recovering from stillbirth: the effects of making and sharing memories on maternal health, *J Reprod Infant Psychol* 31:195–207, 2013.

Engler-Gross A, Goldzweig G, Hasson-Ohayon I, Laor-Maayany R, Braun M: Grief over patients, compassion fatigue, and the role of social acknowledgment among psycho-oncologists, *Psychooncology* 29:493–499, 2020.

Hennegan JM, Henderson J, Redshaw M: Is partners' mental health and well-being affected by holding the baby after stillbirth? Mothers' accounts from a national survey, *J Reprod Infant Psychol* 36:120–131, 2018.

Herziene NVOG-richtlijn 'Zwangerschap tot 24 weken'. 2015. 2020, at https://www.nvog.nl/wp-content/uploads/2018/02/Zwangerschapsafbreking-tot-24-weken-2.0-28-05-2015.pdf).

Ho PC: Development of medical termination of pregnancy: a review, *Bjog* 124:1942–1947, 2017.

Hutti MH, Polivka B, White S, et al: Experiences of Nurses Who Care for Women After Fetal Loss, *J Obstet Gynecol Neonatal Nurs* 45:17–27, 2016.

Johnson OP, Langford RW: A Randomized Trial of a Bereavement Intervention for Pregnancy Loss, *J Obstet Gynecol Neonatal Nurs* 44:492–499, 2015.

Jones K, Robb M, Murphy S, Davies A: New understandings of fathers' experiences of grief and loss following stillbirth and neonatal death: A scoping review, *Midwifery* 79:102531, 2019.

Kamranpour B, Noroozi M, Bahrami M: Psychological experiences of women with pregnancy termination due to fetal anomalies: a qualitative study from the perspective of women, their spouses, and healthcare providers in Iran, *Reprod Health* 17:109, 2020.

Kingdon C, Givens JL, O'Donnell E, Turner M: Seeing and Holding Baby: Systematic Review of Clinical Management and Parental Outcomes After Stillbirth, *Birth* 42:206–218, 2015.

LeDuff LD 3rd, Bradshaw WT, Blake SM: Transitional Objects to Faciliate Grieving Following Perinatal Loss, *Adv Neonatal Care* 17:347–353, 2017.

Lemmers M, Verschoor MA, Hooker AB, et al: Dilatation and curettage increases the risk of subsequent preterm birth: a systematic review and meta-analysis, *Hum Reprod* 31:34–45, 2016.

Limbo R, Kobler K: The tie that binds: relationships in perinatal bereavement, *MCN Am J Matern Child Nurs* 35:316–321, 2010. quiz 21-3.

Lisy K, Peters MD, Riitano D, Jordan Z, Aromataris E: Provision of Meaningful Care at Diagnosis, Birth, and after Stillbirth: A Qualitative Synthesis of Parents' Experiences, *Birth* 43:6–19, 2016.

Lizcano Pabón LDM, Moreno Fergusson ME, Palacios AM: Experience of Perinatal Death From the Father's Perspective, *Nurs Res* 68:E1–E9, 2019.

O'Connell O, Meaney S, O'Donoghue K: Caring for parents at the time of stillbirth: How can we do better? *Women Birth* 29:345–349, 2016.

Peters MD, Lisy K, Riitano D, Jordan Z, Aromataris E: Providing meaningful care for families experiencing stillbirth: a meta-synthesis of qualitative evidence, *J Perinatol* 36:3–9, 2016.

Ramirez FD, Bogetz JF, Kufeld M, Yee LM: Professional Bereavement Photography in the Setting of Perinatal Loss: A Qualitative Analysis, *Glob Pediatr Health* 6:2333794X19854941, 2019.

Ravaldi C, Levi M, Angeli E, et al: Stillbirth and perinatal care: Are professionals trained to address parents' needs? *Midwifery* 64:53–59, 2018.
Richardson A, Ormond KE: *Ethical considerations in prenatal testing: Genomic testing and medical uncertainty, Seminars in Fetal and Neonatal Medicine*, 2018, Elsevier, pp 1–6.
Ryninks K, Roberts-Collins C, McKenzie-McHarg K, Horsch A: Mothers' experience of their contact with their stillborn infant: an interpretative phenomenological analysis, *BMC Pregnancy Childbirth* 14:203, 2014.
Sandelowski M, Barroso J: The travesty of choosing after positive prenatal diagnosis, *Journal of Obstetric Gynecologic & Neonatal Nursing* 34:307–318, 2005.
Thornton R, Nicholson P, Harms L: Scoping Review of Memory Making in Bereavement Care for Parents After the Death of a Newborn, *J Obstet Gynecol Neonatal Nurs* 48:351–360, 2019.
Tiemens-van Putten I: De watermethode: Intacte preservatie van jonge overleden foetus, *Nederlands Tijdschrift voor Geneeskunde* 128:403–406, 2015.
Tseng YF, Cheng HR, Chen YP, Yang SF, Cheng PT: Grief reactions of couples to perinatal loss: A one-year prospective follow-up, *J Clin Nurs* 26:5133–5142, 2017.
Werner-Lin A, McCoyd JL, Bernhardt BA: Balancing Genetics (Science) and Counseling (Art) in Prenatal Chromosomal Microarray Testing, *J Genet Couns* 25:855–867, 2016.
2020, at https://www.loc.gov/law/help/abortion-legislation/europe.php.

CHAPTER 9

Grief

Lara Ras[1], Frederike Dekkers[2], Eva Pajkrt[3], Liesbeth van Leeuwen[3], Jane Fisher[4], Iris Emons[1], Sam Riedijk[5], Iris Jansen-Bakkeren[5] and Karin E.M. Diderich[5]

[1]Experience expert, The Netherlands
[2]Psychotherapy, Private Practice, Rotterdam, The Netherlands
[3]Department of Obstetrics and Gynaecology, Amsterdam UMC, University of Amsterdam, Amsterdam, The Netherlands
[4]Antenatal Results and Choices (ARC), London, United Kingdom
[5]Department of Clinical Genetics, Erasmus Medical Centre, Rotterdam, The Netherlands

Lara: Mae

When I was asking Lara to write about her grieving experience she tried very hard to put it on paper, only to find a powerful writer's block. When we inspected this, it turned out that writing about her grief was causing her a lot of pain. I asked her permission to write about her grief, based on the sessions that we had together and based on the experiences that she shares each year with our students. I cannot write from her perspective but she gladly consented to this approach.

After the welcome and farewell ceremony, Lara and Hugo went to the crematory with their close family. She told the students last year that as she was sitting in the car with the straw basket holding Mae on her lap, she was feeling an unimaginable pain. When the car came to a halt, her body resisted getting up, leaving the car. She was intensely aware that it meant having to let go of the basket, of Mae, leaving there without Mae, and never seeing her again. At this point, Lara and most of the students were in tears.

The period after the farewell ceremony Lara felt numbed, angry; she felt it was unfair that a good citizen who did everything right should have to go through this. While so many irresponsible people, like pregnant women who smoke or drink, get to give birth to healthy children. She struggled to understand what she was feeling. She felt numb, drained, and did not want to go out and meet with people or go to work. She felt very safe with Hugo, who was there for her. She felt that the loss of Mae intensified their relationship.

She also assumed that one day she would become herself again. A lot of things happened. One major event was that after a number of genetic analyses; finally, the cause of Mae's anomalies was identified. It was not campomelic dysplasia, but Antley–Bixler syndrome. This was Lara's biggest fear coming true. Already during the

Prenatal Genetic Counseling
DOI: https://doi.org/10.1016/B978-0-12-823329-0.00009-X

decision-making process, she doubted whether she would be able to live with her decision if it were based on the wrong diagnosis and that is exactly what happened. Lara needed to redo her entire decision process, together with Hugo, to ascertain whether ending the pregnancy was also a decision they would likely have made for this condition. Until this day, Lara is not sure, although she does not regret her decision.

Not only was the actual diagnosis a different one, but Lara also received an incidental finding about herself: she had (mosaic) Turner syndrome. These two test results caused Lara to find herself back where she started in her grieving process twice. She was getting ready to return to her work at school. Instead, she missed work for another school year, trying to recover from all of a sudden being a patient herself, at risk of losing her fertility prematurely. She received screening for her heart as well as for a number of other Turner syndrome–related health risks.

Lara worked hard to recover and started to think about a next pregnancy. Something she feared as much as she desired. She and Hugo decided that they would very much want another child but did not want to pass on Antley–Bixler syndrome again. Preimplantation genetic diagnosis (PGD) was discussed, but as the condition was autosomal recessive and only one mutation was found in Lara, and not in Hugo PGD was not possible.

When Lara started to recover, Hugo came into contact with his emotions. He felt himself increasingly burdened to the extent that his work as an entrepreneur started to suffer. Now that Lara was doing better, he acquired some psychological space for himself and he also needed to grieve.

Lara has come to realize that despite having completed her family, and having grieved, she has never recovered to 100%. She recently told me that she is starting to accept that maybe the old Lara will never come back. The new Lara is less resilient, has less energy to spend in a day, and she wonders if she ever will finish grieving for Mae.

Frederike Dekkers

"Psychological challenge for couples in this time frame"

How do we mourn a child we did not really get to know, other people did not meet, and integrate this loss into our future family life?

How do you mourn the loss of a person who was in your life for only such a short while, but for whom you had such high hopes and dreams?

I have heard many women express their feelings concerning failing not only as a mother but also as a woman and as a wife. They feel they did not succeed at the task women are "made" for.

Daniel Stern proposes in his book "The motherhood constellation" (Innamorati et al., 2010; Stern, 1995) that new mothers create a so-called motherhood constellation. When pregnant, mothers develop a new mental representation about themselves and their future child. A mental representation means that someone is able to symbolize an image of things that they have never experienced, as well as things that do not exist; in this case, it is the mental representation of being a mother (or father). This new mindset, which depends on culture, personality, and own upbringing, forms the so-called motherhood constellation. It develops during pregnancy and may last for many months or even years after birth. To be able to form this mindset, a mother's sense of herself becomes largely organized around the presence of her future baby, the well-being of her child, and their mutual connection. The protection of her child becomes the mother's first and main concern. Daniel Stern proposes four challenges within the constellation. One of the four challenges is for the mother to keep her baby alive, to protect her baby. When couples are confronted with the loss of their child through a termination of pregnancy, the mother did not "succeed" in this task. She was not able to prevent the baby from having anomalies—anomalies so severe that they resulted in a termination of pregnancy. And by "choosing" to terminate the pregnancy, she may even feel like she did the opposite of keeping her baby alive. Losing this child, therefore, does not only start a mourning process, but it also has a heavy influence on the mental image a mother has of herself (Leon, 1992). One of the big challenges in the constellation is not achieved. This may not only lead to self-image problems but can also be seen as one of the causes of why women experience a termination of pregnancy as traumatic, and why this leads to anxiety, depression, post-traumatic stress disorder (PTSD), and complicated grief. In the time frame of aftercare, HCPs should be very attentive and observant how a mother copes with all of this. This can be done by asking how the mother is really doing, how is she coping? Does she eat, sleep, and get up in the morning? And by asking how would the father say the mother is doing? Asking the mother to change perspective can provide us with information about how she is coping. How are they coping together?

We need to know not only for the sake of the mother, but also in the case of a future pregnancy and the relationship that a mother will develop with her baby during and after the next pregnancy.

Mourning in men and women

Women and men deal with a loss such as this differently. Overall, women experience an intense level of grief and show more psychological consequences such as depression, anxiety, and PTSD (Korenromp et al., 2009, 2007; Jones et al., 2019; Kersting et al., 2009; Lafarge et al., 2013; Davies et al., 2005). Mothers have a greater need to talk about their loss, hurt, and mourning (Dekkers et al., 2013). They fight for the existence of their lost child with their surroundings, and they are struggling and learning how to deal with other women being pregnant and seeing a baby (Dekkers et al., 2013). Men seem to take more comfort in distractions such as sports, meeting friends, and sometimes say "I don't know what else there is to say about it. . . ." Jones et al. also found that men reported less intense and enduring levels of psychological consequences than women, and that they show more avoidance and coping such as increased alcohol consumption (Jones et al., 2019). But there is a big task for the father to be able to support his wife post termination (Diamond, 1995). Low-level socioemotional support from partner to mother is found to be one of the significant high-risk factors for a high level of grief 1 year after perinatal death (Tseng et al., 2017). Couples who have lost a baby prior to, during, or shortly after birth often define this loss as highly traumatic and have a higher risk to develop PTSD (Korenromp et al., 2007, 2009 Kersting et al., 2009; Lafarge et al., 2013; Davies et al., 2005; Christiansen, 2017). Couples need to find a way to accept, give room to, and support their different ways of coping. But that is not easy. It is one of the reasons why aftercare is so important.

Precisely 41% of women and 28% of the men sought professional aftercare outside the hospital after the termination of pregnancy (Dekkers, 2013). If couples know that acknowledgment of their lost child is needed to cope with this loss, if they are told by HCPs that many couples seek help from others, trying to integrate this loss into their future lives, they may not feel "incapable, weird, or not strong enough" to handle this themselves. Acknowledgment of the baby's existence and of the couples' suffering needs to be recognized by HCPs and the couples' social network (Lafarge et al., 2014; Jones et al., 2017; Lou et al., 2020).

HCPs should point out that the need for professional psychological help is common. It is not up to us to decide if the case of the couples is "bad enough" to seek counseling, it is up to us to inform couples of the possibilities, and make sure that they know how to receive proper professional psychological aftercare.

Support and network

Parents appreciated it if they were given professional aftercare and missed not having received it as a standard offer from the hospital. Support from a partner, family, and friends in this time frame was valued greatly—as was a memorial service organized by the hospital where the baby was delivered. In this memorial service the names of all the children are read out loud, music is played, and one of the parents who had lost a child tells the story of their loss. Parents find great comfort in being among other parents who have also lost a child, as it allows them a chance to recognize their loss and pain in others. It also helps tremendously in the acknowledgment of their child. Reading books with similar stories also helps parents, and many of such books have been written by parents. In short: people feel less alone when exposed to peers.

Parents missed having been able to talk about coping with their surroundings and the way they deal with the loss of their child. Parents typically are in a life stage in which many friends, brothers, and sisters become pregnant and have babies. And no matter how much they grant others the joy of having a child, they also feel very sad and pained about witnessing other (future) parents' anticipation and happiness. Some even say they cannot bear seeing their loved ones being pregnant or visiting them after birth.

All in all, HCPs and their network should keep a good eye on parents who have been through a pregnancy with prenatal decision-making or a termination of pregnancy. Dare to ask how parents are doing, talk about their loss and grief, and do not avoid it. You cannot hurt parents by asking about their loss—the hurt is already there. Not talking about their loss and their lost child makes parents more alone and tormented.

Eva Pajkrt and Liesbeth van Leeuwen

"Revision appointment"

After termination of pregnancy for congenital anomalies, an appointment is made approximately 6 weeks after delivery. It is important that the time scheduled for this appointment is realistic. In our practice, it is 45 minutes, double the time for a visit after a normal delivery. If there are genetic results pending, sometimes this appointment can be combined with a visit to the geneticist. The appointment is preferably scheduled at the obstetric outpatient clinic with the main care provider, usually the fetal medicine specialist, who also supported decision-making before termination.

The purpose of this visit is to discuss how the couple is doing presently and to stimulate them to talk about their birth experience and how they felt in the weeks thereafter.

To prepare for this visit the care provider should carefully go through the technical aspects of the delivery; did the placenta come spontaneously or was it removed manually? Did the parents give the baby a name? What additional tests were done after birth and are the results there yet? Were the anomalies seen on ultrasound in line with what was described after birth? Or were there additional anomalies? Were there any extra visits after termination because of complaints or complications such as blood loss or fever?

Questions that should follow during this visit are:

Before and during delivery

How did the couple experience decision-making before termination?
Was there enough time?
Was the information sufficient?
How did they experience delivery?
How was the support during labor?
Was the painkilling sufficient?
Did they look at the baby?
Do they have photo's they want to show/share with you?
Do they have suggestions for the improvement of support of labor?

After delivery

Did they bury or cremate the baby or give it away for scientific research?

How was the service?

Whom did they invite?

Weeks after delivery

Have they resumed work?

Is the work provider understanding? No pressure?

Has the woman contacted an occupational physician for resuming work?

How is the relation going together, do they talk about it a lot? How do they deal with the loss? Do they support each other enough?

Are the people in their direct surroundings understanding?

Do they have additional support (social worker, psychologist, etc.) or are they in need for additional support?

Has the bleeding completely stopped? If not is there intermittent blood loss?

Did the woman have her period again?

Future

Desire to have another child?

Recurrence risk?

Organization of care in next pregnancy?

Other suggestions for the conversation

Try to capture the level of emotion that the couple displays in relation to the termination of pregnancy and the loss of their baby, and try to act accordingly. Do not use your own frame of reference. Sometimes a couple can exhibit a lot of grief over a young pregnancy, name a 12-week-old fetus, and organize a whole funeral, while others do not name a 23-week-old baby and leave it behind in the hospital for collective cremation. If the couple has given the baby a name, pronounce the name during the visit in a respectful way to make the visit more personal. We always mention proactively the upcoming due date and address that this may be a difficult time. Parents that recently experienced a termination of pregnancy are usually very conscious of all the pregnant people surrounding them not only in their neighborhood but also in TV series or media. The fact that friends and family members may be pregnant and due around the same time as your patient should be mentioned. This will make the couple feel recognized in their sorrow and grief. After a termination, people surrounding a couple may often have experienced difficulties in understanding the impact of the loss of the baby. Most people only know a miscarriage and a term delivery and nothing in between. On the other hand, parents also reflect that after their loss, numerous people approached them and opened up about their own decision to terminate or a pregnancy loss. Some of these stories may be from a long time ago, which shows the enormous impact. Sharing stories with "experience experts" may be comforting for some. Some hospitals offer support groups; alternatively couples can be referred to existing websites (see later). The opposite is also true, since a grieving couple may have the feeling that they are comforting an "experience expert" instead of the other way around.

Regarding the resumption of work, most of the time the partner has started working again, while the woman has not. There are no guidelines and also no good and bad in this. It just depends on what type of person you are and if you find distraction and joy in your work. Most important is to advise patients not to resume work completely at once, but to build up the hours, and to take time to overcome their grief. Once you have started working full hours again, it is more difficult to say that you do not feel well, instead of taking a small step back if you are still in the process of resuming work.

Usually women do not want contraceptives after the loss of a desired baby. In a Danish cohort of women that had terminated pregnancy between 2000 and 2009 that included 2956 women who had an interpregnancy interval <6 months, the rate of premature birth was 5.6% as compared to 4% in women with an interpregnancy interval >6 months. Women who opt for another pregnancy should take this into account when striving for another pregnancy (Männistö et al., 2017). Most women are considering to get pregnant again as the desire has only grown. Mention that this is common and that here is no need to feel guilty about it. If people need additional support they can be referred to specialized care in hospital, their general practitioner, or to a psychologist. Additionally, around the world, various Facebook groups exist and there are websites offering special care that may provide comfort.

The visit should also include advice on the new pregnancy. Is there a recurrence risk? What should care look like, are additional ultrasounds indicated, or invasive procedures? There should be room for insecurities and acknowledgment that this will be a very exciting pregnancy without any "pink clouds."

Jane Fisher

"Support for the whole family around loss after prenatal diagnosis of fetal anomaly"

At Antenatal Results and Choices (ARC), we recognize that the emotional impact of loss due to termination for fetal anomaly or expected loss after a diagnosis of a fatal condition affects the whole family. While the woman bears the brunt as she is carrying the baby and ultimately has to make the final decision about how to proceed after a prenatal diagnosis, we should not underestimate the psychological needs of the partner. They themselves can neglect their own need to grieve as they concentrate on looking after their partner (Robson, 2002). Men have not always been encouraged to express their feelings or seek emotional support and for some their coping style is to suppress their own needs to be of maximum support to their partner. ARC seeks to address this through its downloadable booklet "Help for Fathers," a forum for men and trained male peer support volunteers. At the time of writing, we are preparing materials to support same-sex partners.

The parents of women and men facing bereavement after prenatal diagnosis can also be profoundly affected. They must watch their offspring go through a painful and often harrowing experience and feel powerless to "make things better" which will often be their first parental instinct. They also have to deal with the loss of their potential grandchild. Furthermore, the more recent technological changes in prenatal diagnosis may be unfamiliar to them. Some "grandparents" contact the ARC helpline to find out what kind of help might be available to their son or daughter. The helpline team will always encourage them to use ARC as a source of support for themselves and point them toward the downloadable PDF on our website, "Help for Grandparents."

Parents who opt to terminate an affected pregnancy often express anxiety around how to talk to their other children, living, or prospective about what happened. The conflicted feelings they have about actively choosing to terminate can sometimes cause them to believe they have to "admit" the truth of what happened and face what they believe will be an inevitable judgment. The ARC booklet, "Talking to Children," prompts them to consider what other children actually need to know. Children do not need all the details (it could be argued that this is part of the parents' story and sharing all may overburden their children). However, at the time of the termination, children will need to be told something age appropriate to explain their parents' distress. If they are aware of the pregnancy, they need to know the younger sibling they were anticipating is not to be.

Iris: Noud

After 6 years, I noticed that I was always scanning my environment, scanning for people who had more than two children. If people had more than two children (which I had after having given birth to my son, Faas), I would condemn that. If people said they had two children, I would inquire whether they desired a third child. And if they did, I would lecture them on the risks of anything going wrong and I would interrogate why they would be willing to take that risk when their two children were healthy. In my opinion, wanting a third child was greedy and ungrateful. I felt angry when people had more than two children, and I just could not get my head around people willing to take the risk of having their life changed forever like I experienced, when in my view "they had it all." I had an amazing life, a wonderful partner, a healthy boy, and fantastic job and great friends. I just wanted what is so common among people with one child: a healthy brother or sister for Mats, and my life changed forever because of it. The scanning became an obsession and I had to do something about it. I needed the help of a psychologist because I was so angry at the entire world but it was so difficult to actually take the step. Would a psychologist really think that my problem was big enough to address? I had my second child, what was I complaining about. I decided to approach the same psychologist I spoke 5 years before about this.

During the sessions, it became clear that I had not grieved properly. I had to grieve my loss, the loss of my son Noud, 6 years after I lost him. But why was I so angry with people wanting or having a third child? I did not even want a third child myself? Why did that trigger me at all? And then came a moment when my psychologist said: "Do you hear what you are saying, Iris? You say that you do not want a third child, but you *have* three children: Mats, Noud, and Faas." And that was it. I have three children. This realization opened a door to reconnect to my grief again. It was intensely painful, and it diminished my obsession.

I wish I would have done things differently. I wish that I would have held him in my arms, that I would have given him a name straight away, I wish that I would have talked about him much more, and that I had not felt ashamed, because for years, shame is what I felt. I had given birth to a baby with multiple congenital anomalies and we decided to end the pregnancy. I tried to perceive that as something small, to simply move on. But that is not something small. It is big. It is huge and something that I carry with me for the rest of my life. I carry Noud in my heart forever, with all my heart. Deep in my heart and that feels so right, nobody can touch that.

What helps according to Iris:

- Aftercare initiated by the hospital. A phone call, a note, a moment of attention for the intense situation you went through.
- Offer advice regarding emotional processing. Recommend to take time for this processing and counsel toward talking to someone who has experience with this.

Sam Riedijk

Giving birth turns the couples into parents, and whether or not the child is alive, and whether or not couples decided to end the pregnancy, it makes no difference. Mostly the mother will be most strongly attached to the child, and the father will be mostly concerned with caring for the mother. Sometimes couples feel as if having made the decision to end the pregnancy deprives them from the right to grieve. However, all couples grieve when they lose a child, regardless of whether the loss was due to the decision to end the pregnancy or a natural cause (Salvesen et al., 1997). Having to carry the weight of this decision does weigh heavily on parents. In congruence with scientific literature (Korenromp et al., 2009), most parents I speak with, even years after, even if they have no regrets, still find carrying the responsibility of this decision to be traumatizing. It is important to recognize that on top of this psychologically normal response, some individuals may require more psychological support to move through their grieving process. Having used antidepressants, being younger or more highly educated are some of the risk factors associated with complicated grieving (Cole et al., 2016). In addition, the choice whether or not to see the baby may be a difficult one for parents, as they often state fearing that they may become overly attached or traumatized by the image of their deceased baby. Sometimes it takes some time to become strong enough or ready to see the deceased baby. In my experience, I have rarely met parents who, in hindsight, regretted seeing their baby, but I have met couples who, in hindsight, regretted not seeing their baby. Understanding that couples may be in doubt is important, giving space to express their ambivalence is also important (Sun et al., 2014). Couples need compassion, sensitive care, and validation of their emotions (Peters et al., 2016). Couples are often reassured by hearing about the experiences of other couples, as it normalizes their responses and elicits trust that they too may come to experience the delivery and period after as something not only sad but also special.

Saying farewell

Parents in our hospital are offered a choice to either take care of the funeral or cremation themselves or leave this up to the hospital. In my experience, roughly as many parents decide to take care of the funeral themselves, whereas the other accepts the offer of the hospital cremation service. A minority of parents leave the hospital and never see their child again, while a minority of parents bring their baby home before the funeral. In most instances the time between birth and farewell is no more than 3 days. Most parents, irrespective of choice, organize a small farewell ceremony with a few of their closest loved ones. It is only after this ceremony that for parents, the grieving for the lost child begins. Grieving ceremonies are psychologically beneficial (Camacho-Ávila et al. 2019).

The first few days, most parents describe feeling closer to each other as partners. They go through child birth together, name their child, invite a few important others to share a moment with their baby with them, and organize the farewell ceremony. To couples, this feels like a safe bubble in which they are synchronized and are truly in it together. As long as the partner stays at home, the bubble remains warm and intact. However, as soon as the partner leaves for work again, the bubble bursts and the parents each starts to write a different chapter in the book of grieving. This relationship dynamics phenomenon often causes tension between parents. It is not even a male–female difference, but in my experience a pregnant–nonpregnant issue, as I have seen the exact same dynamics in lesbian couples. Here, I will call it woman and partner for readability.

Woman

After giving birth, women often describe feeling exhausted and proud and overwhelming love for their child. The time spent looking at the baby and cherishing it are essential for creating memories that mothers will hold onto tightly the rest of their lives. Any pictures taken in this moment are also important for creating a footprint for the baby. It is important that she codesigns a farewell for her baby that suits her, that does justice to her motherhood. It is the only way in which she can care for her baby.

It is good to encourage parents to each contribute to the farewell ceremony, to make it their own, as it has the potential of comforting them in their most difficult grieving moments. A farewell that feels good, that was made their own, is something that couples will look back on in acceptance.

When the moment of farewell has passed, women describe feeling numb for a few weeks. They wake up thinking it was a dream. They find themselves talking about what happened as if they described something that happened to someone else. The reality of the loss takes some time to surface, and it tends to wash over them in waves.

Many of the women I spoke expected their grief to follow a linear curve. Starting at the bottom and then gradually climbing back to normalcy. What they experience is actually much more comparable to waves or labor contractions. Pain that suddenly sets in, intensifies, peaks, and then resides. Waves of sadness, broken-heartedness, anger, and despair were alternated with moments of calm, of feeling okay, and of feeling guilty for not experiencing grief. In my counseling sessions, I always find myself explaining this pattern to couples, or women, and see it resonate. The recognition of this experience is comforting and provides reassurance and a frame of reference. When parents seem to be in doubt of what they are experiencing or when they are trying to resist, be strong, or find distraction, it is extremely useful to provide an explanation for what they are feeling and to validate that what they are experiencing is normal. For most parents, this is their first encounter with intense grief. Normalizing can be done in many words. Mine tends to be something like:

> *Most people I have met who were in a situation similar to yours have told me they felt exhausted, they did not recognize themselves, in the sense that they did not enjoy what they normally enjoyed, they did not care for things they would usually find important, their bodies felt different, they could not concentrate, or make decisions. Deciding which yoghurt to choose in the supermarket was impossible. People would come home from shopping and put their keys in the refrigerator and the cheese in the closet. Even sleep is affected. I am wondering what it is like in your situation?*

Every time people express relief, to hear that this is normal and that they are not losing their sanity. Such small efforts to offer some frame of reference actually provide a lot of comfort.

It seems that grieving is something that happens in each cell of your body, each piece of your psyche.

I have noticed that women grieving by cocooning seem to recover more quickly from their loss. Some women make their world very small, stay at home, only let those people in who feel safe and bring something that benefits her. These women seem in close contact with their pain and describe in some sense embracing their grief. They cry when it happens, they laugh when they feel it, they know that work is for later, and they spend their time on the couch, writing down their memories, creating a photo album, or choosing a symbolic piece of jewelry. In a symbolic way, they are exerting their motherhood, creating existential right for their baby, creating a footprint.

Sometimes women feel that they should not give into their emotions, feeling that they should seek distraction and be strong. It may be that I have a bias in my experience because only those women come to me for whom this does not work. But in my experience, it was often these women who had sleeping problems or who experienced delayed grieving, sometimes up to years later. I am not judging this, because I think that it can take time to get ready for grieving. I do think that all grieving needs energy and attention. Distraction can help to recharge and it can also take away the energy and mental capacity that is necessary to grieve.

Grief is invisible though and, therefore, can be difficult to acknowledge or even recognize. Using metaphors may provide a powerful tool to visualize grief. The metaphor that I like to use is one of a broken leg.

> *Imagine your leg is broken in three places. To you it would probably seem entirely valid to sit on the couch all day and rest, to have others care for you for a while. Of course at some point your leg will need to learn to move again, starting with some small exercises and gradually building up, adding more burden to your leg. In fact your grief is something like a severe wound to your psyche and requires the same care as your leg. If your leg was broken in three places, you would probably not train for the marathon. Likewise, you should not ask yourself to get back to work and function like you are used too when you are grieving.*

Depending on the counselor and the person in front of you, many other metaphors can help people get grip on what happens to them.

On average, most women I have encountered tended not to feel ready for work for around 4 months after birth (Geerinck-Vercammen and Kanhai, 2003). However, my colleagues in fetal medicine indicate that they consider slowly returning back to work anywhere between 2 and 6 weeks after delivery as normal.

Often women have conveyed that the due date is an intensely painful moment. Only after the due date the true loss "starts," because this is the date that the baby should have been there, and the discrepancy between what is and what should have been becomes tangible. The period before the due date is more of a twilight zone. This also means that all the first times after the loss start at the due date; the first summer holiday, first Christmas, your birthday, and other meaningful dates. Many women I have met told me that somehow in their minds their baby was growing up, one birthday at a time. In that sense, I see how the relationship these mothers have with their deceased child changes over time but never ends (Keirse, 2017).

Sometimes women wonder what is an appropriate time to get back to work. The answer has proven to be quite simple: when you feel like it. In my experience, women gradually enlarge their cocoon and increasingly feel like going out and letting the world back in, at their own pace. They will know they are ready because they get curious about how their colleagues are doing, they feel like contributing again. They feel that they can offer support to a friend in need again. This inner compass is very valuable. Feeling ready helps to face the challenges of the outer world, like facing women who are pregnant or had their baby around your due date. I have yet to meet the woman who has discovered how not to experience pain when confronted with other people's pregnancies and baby happiness.

What is important for women is to find a way in which their baby has a place in their lives, a name that is mentioned not only by her but also by her loved ones. Gradually they develop their own ways of answering the question: When will you have a baby? Or how many children do you have? Questions that, during the first period of grieving, are extremely painful, after a while become uncomfortable, and eventually become part of life without causing imbalance anymore.

Partner

The partner has a different task in the grieving process. Throughout the pregnancy, he has become increasingly involved in caring for the pregnant woman. He has witnessed her changing body, her moods, what is on her mind. He has been there when the news devastated her and him, and he has felt the responsibility to be strong, to be there for her, to stay rational where she became emotional, and to manage appointments while she was struggling with feeling her baby kicks in her belly in the midst of bad news, in the roller coaster of hope and fear. When they decided to end the pregnancy, he probably believed that was the best thing to do for the baby, them, and their family. However, he also felt overwhelmed by her courage and his guilt that she had to give birth and he felt powerless when he was standing by while she was in pain. Once he sees his baby, he too feels pride and love, but often even more so of his wife. When she is grieving, he feels the responsibility to keep the world spinning. In this phase, woman and partner are writing different chapters in the same book, each fulfilling important tasks, but not the same tasks.

Therefore partners often resume work quite rapidly, ranging between 2 days and 3 weeks after birth. The woman tends to experience a great deal of anxiety when she is left home alone. This is the moment the safe bubble bursts and some tension rises between partner and woman.

Pregnant–nonpregnant differences

When the partner ensures that the world keeps on spinning, he might discover that he and his wife have conflicting needs (Ramdaney et al., 2015). While he needs distraction, she needs to introspect, while he needs action, she craves peace, he wants to meet his friends, while she is too anxious to discover the next friend has fallen pregnant. Often partners find themselves offering each other the support they need themselves, only to find out that this does not match the need of their partner at all. Without knowing anything about the couple in front of me, I have witnessed the relief of recognition when I tell them what often happens between couples during this phase.

> *Often the partner will say to the woman that she needs to get off the couch, leave the house, meet with friends, seek distraction. He does not realize that this is the last thing on earth she wants. Likewise, women often ask their partners how they are feeling, inviting them to introspect and talk about their emotions. She does not realize that he will have a hard time connecting with his pain, because he is creating space for hers. If both partners would give each other what the other person needed it would be something like this. She would tell him that he should definitely go out, get on his bicycle or meet up with friends, and that she will be fine in the meantime. He would be offering her support by sitting next to her, offer his shoulder and ask her to tell him what's going through her mind. Embrace her, allow her time to repeat herself, and not offer any solutions. He is the solution. Her pain can dissolve in his attention, even if it is just for a moment.*

"I can do that," is a frequent response. Understanding that both have different needs helps to build a bridge. Additionally, when the household takes over, and couples resume their normal daily rhythm, spending time together without a phone in the hand, a TV in the background, or other distractions help couples to bridge the differences in longer term grieving.

Iris Jansen-Bakkeren

Guiding siblings through the termination of pregnancy

Often pregnant couples already have children during the pregnancy in which the abnormalities are detected. Depending on the age of the child, they are aware that their mother is carrying a baby. They know that the pregnancy will result in a little brother or sister. When expecting couples receive the devastating news that abnormalities are detected in their unborn child, and they go through the roller coaster described earlier, this also affects their other children. These children, irrespective of their age, sense that something is going on and that their mom and dad are sad.

A lot of parents struggle with how to inform their children about the abnormalities in the pregnancy (Avelin et al., 2011). This is especially the case when couples are full amidst the decision-making process themselves. What we have seen, however, is that it is important to inform children about what is going on and why their parents are sad. If children are not informed and stay in the dark, they might start to blame themselves (Capitulo, 2005). They may believe that they are the reason their parents are sad. It sometimes seems that children, especially when they are under the age of five, resonate the feelings of their parents. In this case, children seem to be mirroring their parents' inner turmoil.

Depending on the age of the child, the couple can inform them that something is wrong with the baby or the baby is sick. Parents can explain that this makes them very sad, and that is why they are acting different.

When couples have made the decision to end the pregnancy, their other children should also be informed about this (France et al., 2013). We have experienced that including children in this process makes them understand better and helps children in their own grieving process. The effect of involving the child is that it stays part of the inner circle, where it is usually safest. When parents try to protect children by not involving them, they feel excluded. Being excluded is being out of the safe zone. When children sense tension and emotional turmoil, being excluded may be even more distressing. It helps to realize that young children may feel less burdened by the loss of their sibling, as they do not yet possess the cognitive capacities to feel the pain of missing "what should have been." They experience what is, and for them there is no painful discrepancy with what should have been.

Depending on the age of the child, parents can explain that the baby is "broken" and cannot live or that the baby would be in a lot of pain if it would be born alive. When children are a little older, parents can explain that they give birth to the baby so that the baby will not have to suffer. In all cases, it is important to stress to children that the unborn child is severely ill. Otherwise, children may become deeply troubled when they have the flu or fear they might die when they fall ill.

While parents lose their child, children lose a brother or sister. Children grieve the loss of this brother or sister in their own age-appropriate way (Kappeyne van de Coppello, 2016; Keirse, 2009). Therefore it is important to include the children in the process of saying farewell to the deceased child. We hear from a lot of couples that they struggle how to do this, that they worry it might be traumatizing for their child (ren). From what we learned from parents' experiences in our clinic is that children took the situation with much more "lightness" than their parents. We have yet to encounter the child who was traumatized by being included in the farewell ceremony (Keirse, 2017). We must note here that in these families, there typically was no psychopathology or other problematic family dynamics issues.

Including children allows them the possibility to mourn the loss of their brother or sister. Ways of including the other children, depending on their age, are having the child see or hold their deceased brother/sister, having the child watch photos of their brother/sister, giving the child a role in the farewell ceremony, having the child make drawings for their brother/sister, or having the child buy a gift/stuffed animal for their brother/sister. Couples may feel anxious about talking about their deceased child or show their emotions when their other children are present. Most parents are aware of their responsibility as caregivers and do not want to stimulate their children to comfort them in their grief, as this would cause role reversal. Indeed, parentification has a negative impact on the psychological well-being and development of children (Onderwaater, 2014). However, allowing your child to hug you and give you a kiss to comfort you in your sadness does not equal role reversal. I often tell parents that there is a boundary indeed, if parents were to actually rely on their child to feel better and seek out their child to receive comfort, it would be crossing the boundary. But allowing your child to practice with comforting behavior that they themselves receive from their parents is adequate parenting. Most parents recognize this boundary when we discuss it this way and feel reassured.

Being transparent about your emotions when you feel overwhelmed sets the example for the children that they are also allowed to feel sad/confused/angry and that it is normal to talk about their brother/sister if they want to.

During these difficult family times, it is, however, important for children to maintain their normal routines as much as possible (Keirse, 2009). Attending school, practicing sports, playdates, etc. are beneficial to help children in their grieving process. As children spend a lot of time in school, informing their teacher ensures that children also feel safe to talk about what is happening/what happened in the class. The parents could discuss with their child and the teacher what they would like to tell in the class about their brother/sister. Furthermore, the teacher could observe how the child is doing and whether additional support such as a school counselor or child psychologist would be necessary.

Children grieve different than adults do. Children can be very sad, wanting to snuggle up with mom and dad and want to talk about their brother/sister. The next moment, minutes, an hour or hours later they can be happy and joyful again and want to continue their daily activities such as playing or school (Kappeyne van de Coppello, 2016; Keirse, 2009). This does not mean they do not mourn the loss of their brother/sister; this is just a child's way of keeping it maintainable for him/herself. For parents it is important to stay available for their children, for their questions, concerns, and tears. By speaking about the deceased child among the other children, the children learn that it is okay to talk about him/her and that their brother/sister is part of the family.

Children develop quickly and learn more about life and death in school when they grow older. Therefore in each developmental stage, children will be able to perceive the death of their sibling from a new perspective (Kappeyne van de Coppello, 2016; Keirse, 2009). Based on their growing knowledge and development children keep asking new questions, using the new insights they gathered. These questions remain for years after the termination of pregnancy has occurred. Some parents worry when in a new developmental phase their child is all of a sudden preoccupied with the loss of their sibling again. It is, however, a psychologically adequate reaction for children to attribute new meaning to their loss with their newly developed capacities. When parents understand this, they indicate actually liking these questions, because it shows that for their other children the child they have lost is also part of their life. By staying mentally available for their children, parents can guide and support their children through the grieving process.

Karin Diderich

When we discuss the termination of pregnancy, we also discuss the different ways of saying goodbye. We advise to include previous children in (part of) these rituals. They can draw a picture for their little brother or sister, they can come by in the hospital with, for example, a grandparent and say goodbye if they want to. Once I entered a room for a physical examination of the fetus and found the family gathered around the 4-year-old big sister who was cradling the fetus in her lap, a wonderful moment. In contrast, I have heard back of people who did not include their daughter in saying goodbye who found that the daughter was terrified for the empty nursery and had scary images in her head. Sometimes a couple has already decided—with the best interest of the child in mind—not to include the child in any of the rituals. I remember one father who was also a teacher and had a very strong conviction that it would be harmful for his child to include him. Having (read about) different experiences myself, I persevered in telling him about this which actually agitated him so much that I was in doubt whether that had been wise. When I saw the couple again, he told me that in the end they had included their son and that was the best thing to do. Sometimes, a couple feels their child is too young to understand any of it and thinks it is better not to include the child. I tell them that children sense much more than we as parents might comprehend. When a couple does not quite see that, I sometimes share an experience of my own.

> *My own mother died of cancer at a young age and a very good friend of mine was terminally ill with cancer. I used to visit her with my three-year old daughter to help out with laundry etc. At one point the nursery teachers asked me to come by at the day care center because my daughter was not her real self in their eyes. We sat down talking and came to the conclusion that this was most obvious the day after my daughter and I had visited my friend. I sat down with my daughter and it appeared that she had come to the conclusion that mothers, including her own, die at a young age. I told her I was not sick and from then on I informed the nursery teachers every time we had visited my friend and my daughter became her real self again.*

In a short time, I can, thus, illustrate how much children can sense and imagine. Sharing the true situation with their parents seems to be the less harmful option. In addition, I offer couples an appointment with the psychologist who has far more experience in this field.

Main recommendations

- Dare to ask how couples are feeling and coping.
- Refer couples in time when you feel they are not coping well.
- Be attentive to anxiety, depression, PTSD, and complicated grief symptoms.
- Point out the high percentages of received professional psychosocial aftercare; it is common for couples to need psychosocial care.
- Actively acknowledge the loss of parents and give room to mourning.
- Advise parents to include their other children in the farewell of their brother/sister.
- Discuss the wish for a new pregnancy.

References

Avelin P, Erlandsson K, Hildingsson I, Rådestad I: Swedish parents' experiences of parenthood and the need for support to siblings when a baby is stillborn, *Birth* 38:150–158, 2011.

Camacho-Ávila M, Fernández-Sola C, Jiménez-López FR, et al: Experience of parents who have suffered a perinatal death in two Spanish hospitals: a qualitative study, *BMC Pregnancy Childbirth* 19:512, 2019.

Capitulo KL: Evidence for healing interventions with perinatal bereavement, *MCN Am J Matern Child Nurs* 30:389–396, 2005.

Christiansen DM: Posttraumatic stress disorder in parents following infant death: A systematic review, *Clinical Psychology Rreview* 2017:60–74, 2017.

Cole JC, Moldenhauer JS, Berger K, et al: Identifying expectant parents at risk for psychological distress in response to a confirmed fetal abnormality, *Arch Womens Ment Health* 19:443–453, 2016.

Davies V, Gledhill J, McFadyen A, Whitlow B, Economides D: Psychological outcome in women undergoing termination of pregnancy for ultrasound-detected fetal anomaly in the first and second trimesters: a pilot study, *Ultrasound Obstet Gynecol* 25:389–392, 2005.

Diamond MJ: Someone to watch over me: the father as the original protector of the mother-infant dyad, *Psychoanalysis and Psychotherapy* 12:89–102, 1995.

France EF, Hunt K, Ziebland S, Wyke S: What parents say about disclosing the end of their pregnancy due to fetal abnormality, *Midwifery* 29:24–32, 2013.

Innamorati M, Sarracino D, Dazzi N: Motherhood constellation and representational change in pregnancy, *Infant Ment Health J* 31:379–396, 2010.

Kappeyne van de Coppello CJ: Kinderen in rouw: aandachtspunten bij diagnostiek en behandeling, *Kinder- & Jeugdpsychiatrie* 43:84–91, 2016.

Keirse M: *Helpen bij verlies en verdriet*, 8th ed., Tielt, 2017, Lannoo.

Keirse M: *Kinderen helpen bij verlies: een boek voor wie al van kinderen houdt*, 1st ed., Nederland, 2009, Lannoo.

Kersting A, Kroker K, Steinhard J, et al: Psychological impact on women after second and third trimester termination of pregnancy due to fetal anomalies versus women after preterm birth–a 14-month follow up study, *Arch Womens Ment Health* 12:193–201, 2009.

Korenromp MJ, Page-Christiaens GC, van den Bout J, et al: A prospective study on parental coping 4 months after termination of pregnancy for fetal anomalies, *Prenat Diagn* 27:709–716, 2007.

Lafarge C, Mitchell K, Fox P: Perinatal grief following a termination of pregnancy for foetal abnormality: the impact of coping strategies, *Prenat Diagn* 33:1173–1182, 2013.

Leon IG: The psychoanalytical conceptualization of perinatal loss: A multidimensional model, *Am J Psychiat* 149:1464–1472, 1992.

Männistö J, Bloigu A, Mentula M, Gissler M, Heikinheimo O, Niinimäki M: Interpregnancy Interval After Termination of Pregnancy and the Risks of Adverse Outcomes in Subsequent Birth, *Obstet Gynecol* 129:347–354, 2017.

Onderwaater A: *Rouw vanuit systeem perspectief, Handboek rouw, rouwbegeleiding, rouwtherapie*, 1st ed., Nederland, 2014, Witsand uitgevers, pp 219–231.

Robson FM: 'Yes!–a chance to tell my side of the story': a case study of a male partner of a woman undergoing termination of pregnancy for foetal abnormality, *J Health Psychol* 7:183–193, 2002.

Salvesen KA, Oyen L, Schmidt N, Malt UF, Eik-Nes SH: Comparison of long-term psychological responses of women after pregnancy termination due to fetal anomalies and after perinatal loss, *Ultrasound Obstet Gynecol* 9:80–85, 1997.

Stern DN: *The motherhood constellation: a unified view of parent-infant psychotherapy*, 1995, Taylor & Francis Ltd.

Sun JC, Rei W, Sheu SJ: Seeing or not seeing: Taiwan's parents' experiences during stillbirth, *Int J Nurs Stud* 51:1153–1159, 2014.

CHAPTER 10

Building a life again

Sam Riedijk[1], Frederike Dekkers[2], Iris Emons[3] and Lara Ras[3]
[1]Clinical Genetics Erasmus Medical Centre, Rotterdam, The Netherlands
[2]Psychotherapy, Private Practice, Rotterdam, The Netherlands
[3]Experience expert, The Netherlands

Lara: Mae

When both were recovered sufficiently, Lara and Hugo decided to take their chances and try for another pregnancy. Lara was sick throughout the entire pregnancy and was admitted to hospital on several occasions to be treated for dehydration. During her subsequent pregnancies, Lara experienced difficulties in feeling a connection with the baby in her belly. Because of what she had experienced, she did not feel confident in her intuition. She tried to ignore her pregnancies for as long as possible, which proved to be quite challenging considering her nausea. Ignoring her pregnancies made her experience guilt feelings. It made her doubt whether she would be able to truly love this baby, and she feared her baby would sense her psychological detachment and feel rejected and hurt by it. Although the baby was deeply welcomed, Lara's heart was so damaged that she feared not being able to deal with a loss like this ever again. She was only really able to embrace both subsequent pregnancies after favorable test results from amniocentesis and the 20-week ultrasound (US). These test results would mark a turning point after which the nausea would decrease. The true relief, both times, only happened after delivery of a healthy baby boy, and 3 years later a healthy baby girl. Both times postnatal attachment was uncomplicated.

Hugo and Lara have handmade pottery, that they made themselves, in which they keep the ashes of Mae. These pots stand in a nice spot in their living room. As such, Mae is part of their daily lives. And they often tell their children about their big sister Mae.

Frederike Dekkers

Many couples become pregnant again after a loss. With a loss in their history a new pregnancy will never be the same. Campbell-Jackson et al. found that mothers and partners experience high levels of anxiety about their baby's well-being during a next pregnancy, and that partners report a lack of opportunity for grieving (Campbell-Jackson et al., 2014). Partners tend to take care of the mothers and are vulnerable to not taking care of themselves.

Prenatal Genetic Counseling
DOI: https://doi.org/10.1016/B978-0-12-823329-0.00005-2

Case P

A couple went through an uncomplicated (first) pregnancy in which they had a minimum of diagnostic evaluation. There were no anomalies known in both families and so the couple decided against the 20-week US. The pregnancy went well. But after birth it was immediately visible, the baby boy was severely affected. Trisomy 13 was diagnosed and the boy died 5 days after birth.

The couple was totally unprepared for this and felt overwhelmed and devastated.

After 2 years of mourning and much doubt about having another child, they decided to try to become pregnant again. The mother felt insecure, feeling guilty toward her son. She felt she was trying to replace him for a healthy version. When the mother was pregnant, she could not really feel happy, anxious about what they would possibly be facing. The father felt more positive. This time they did all the diagnostic evaluation that was available. No anomalies were seen, but it was hard for the mother to attach to the new baby that was coming. She could not feel reassured. The mother somehow felt relieved that this time, the baby was a girl. But the movement she felt, felt the same as in her first pregnancy, reminding her about her lost son. At 30 weeks, they started to clear out a room, but it needed to be a different room than the nursery for their first child. At 36 weeks the mother became more agitated and more anxious, fearing another overwhelming experience at birth. She had a hard time focusing on the good and healthy things in her pregnancy. The delivery went well and the mother gave birth to a blazing baby girl. The mother felt like she did not know the girl, she felt strange. The midwife, who knew about the history of the couple, visited the family every 2 days for 2 weeks and saw little progress in the mother's attitude toward her daughter. The mother mostly left the care to the father, who was very excited. The midwife shared her observations, explained that this can happen after the loss of a child, and that the mother was not the only one who had difficulty bonding with a new baby after such a loss. She offered to refer her to an infant mental health specialist to help increase and strengthen the bond between mother and daughter. The mother started to cry and said she felt relieved. She thought that she had no more love left for this child but was also afraid to attach. She had learned that a child could "just die."

The couple went for therapy, for the mother mostly to learn how to enjoy the relationship with her daughter—to bond but also to trust. For the father, it was mostly to support his wife and help her enjoy the contact with her daughter.

What can we learn?

This midwife had a very good antenna for noticing that the bond between mother and daughter was not developing naturally. She observed them well and was able to discuss her observations in a way that made the mother feel validated and seen. She deserves a great compliment for noticing, discussing, and referring this family to the right kind of care, in this case an infant mental health specialist. She has prevented this family from a potential insecure attachment between mother and daughter, which, for the child, could have led to anxiety, difficulties regulating emotions, and interacting competently with peers, which may lead to further anxiety. And for the mother, an insecure attachment can lead to more feelings of insecurity, feelings of incompetence as a mother—all on top of having experienced losing a child.

It is advisable to ask during a new pregnancy how the quality of bonding is coming along. And if signs of difficulty in bonding are observed during the pregnancy, it is very important to discuss these observations. Couples can be supported and guided throughout the pregnancy and be prepared as best as possible.

The conclusion is that discussing your observations in a validating, acknowledging, and empathic way and knowing where to refer couples for support is of great help.

Attaching to a new baby while still mourning a lost one is hard. Markin studied the difficulties bonding to the unborn baby during pregnancy after a prior loss of pregnancy (Markin, 2018). This is important for us as healthcare professionals (HCPs), because the quality of bonding during pregnancy is predictive for problems in the mother–infant attachment relationship. There is a bigger chance of an insecure attachment for babies born after loss. An insecure attachment itself may contribute to anxiety, but insecurely attached children also are more likely to have difficulties regulating emotions and interacting competently with peers, which may further contribute to anxiety (Kerns and Brumariu, 2014). An insecure attachment can also be caused by a term "ghost in the nursery"—a concept Selma Fraiberg spent her life's work on (Fraiberg et al., 1975). After the loss of a child, couples can take the burden of their loss into the next pregnancy, not being able to really connect to the new baby, out of grief. With every new check or US, couples may think about their lost child. They may protect themselves by not attaching to this new baby, to avoid potential new hurt. Being pregnant again is the biggest reminder of what couples have lost. The lost child becomes a "ghost in the nursery" that affects the future bond with the next child.

O'Leary also studied grief after a lost child and the impact on prenatal attachment in the subsequent pregnancies (O'Leary, 2004). She suggests that a new layer of grief over the loss of a previous child is formed when a mother becomes pregnant again. This grief needs to be recognized because it can lead to pathology such as anxiety, depression, or complicated grief. The loss needs to become a part of the autobiographical story of couples, and many couples need help with this.

Others often do not see this loss as a family tragedy and do not think of the baby as a real family member. Especially with a lost firstborn, people in the couples' social environment easily overlook that couples did become parents but have no child to care for. Women who have lost a child face a significantly higher risk of anxiety and depression in a subsequent pregnancy (Gravensteen et al., 2018). Grief is less likely to become "pathological" when others acknowledge the couples' experience of their loss and give room to parents to mourn. To be able to mourn the loss of a child, the existence of this specific child must be acknowledged. Acknowledgment of his/her existence means that people ask how the baby looked, bear to watch photographs, ask couples how they feel, ask how much they miss their child not being in their lives. Not once, but once in a while for as long as couples need—taking time to remember the lost child together with the couples.

Parents often do not receive guidance in a next pregnancy (Dekkers et al., 2019). In my opinion, discussing the option of seeking professional psychosocial care during a next pregnancy after pregnancy termination or child loss just after birth should be standard care. The intensity of this aftercare, of course, is done in consultation with couples. Specific attention should be paid to addressing the fear of having another abnormal pregnancy. We should make couples aware of their possible avoidance of attachment to the new child for fear of another loss. Couples often feel that they have to keep on grieving for their lost child out of loyalty; they feel guilty about happy moments, or for not thinking about their lost child for a period of time. Some feel as if they are replacing their lost child by having another child. We need to help couples attach emotionally to this new child that is coming into their lives, alongside their thoughts and feelings about the lost child. They will still grieve and never forget their lost child, but they need to feel "permitted" to want and to have another pregnancy.

In my experience the advice couples are given as to how long they should wait to get pregnant after a loss highly depends on the attending gynecologist. Women reflect that this was difficult for them at times. Couples in this situation desperately want to do the right thing and avoid unnecessary risks. Thus they are keen to follow advice from HCPs, even when an advice is fairly neutral. But for some couples, that advice meant that a friend could get pregnant faster than they did, which, in turn, reinforced the parent's grief and feelings of loss. If it is medically justified, HCPs should consult with couples as to when a next pregnancy is possible. And if HCPs have doubts about the mental state of couples, or if they feel that a new pregnancy would come too soon, they should collaborate with a member of the psychosocial care team to explore the couples' wish.

Iris: Noud

We had an appointment at the clinical genetics department for 1.5 months later. The gynecologist from hospital number 1 said during our closing conversation that it was not necessary to go to this appointment. It just had not been a good pregnancy and with one healthy son, we should not make it bigger than it was and just try again. There was no recurrence risk.

Still, the appointment was scheduled and we thought; let us go because we still did not know what had gone wrong with our baby. We met with a clinical geneticist, and she said right away, I really think this is a syndrome and there could be a 25% recurrence risk with both of you being carriers of something. You can imagine we both were perplexed and I immediately became emotional. Knowing that I would get pregnant again in the future had pulled me through these difficult weeks, and now we were told that there was a significant chance of recurrence. The clinical geneticist then asked, have you ever spoken to a professional about this? And I had not. In the entire process, no one had ever told me that I could get help for this. Everyone said move on, this was bad luck and just try again. And then I started therapy with a medical psychologist. And she let me tell my story every time. My worst thoughts, my fears for the future, would I ever have that second child. And she also asked me why I did not name my son, because I always talked about "the baby." And "the baby" slowly grew into "Noud" because that was the name we already had in mind. Noud!

The clinical geneticist did a lot of tests for different syndromes, but each time she came back with the result that it was not that. After 6 months, we got the green light from her to try to get pregnant again. Until now, we still do not know which syndrome Noud had, and the clinical geneticist has had tests done with new techniques up to 3 years later.

I was pregnant exactly 1 year after Noud was born. We kept this pregnancy to ourselves for a long time because we were so afraid it would go wrong again. The 20-week US was terribly scary, but luckily everything turned out to be fine there. But even after the US, I could hardly trust a good outcome and the fear stayed with me for 38 weeks. At the hospital, they were very understanding, and fortunately, when I felt uneasy, I could have an extra US. Then I got to choose whether I wanted a cesarean section or give birth naturally. My first son was born through a cesarean section and Noud "naturally." I wanted to experience what it was like to naturally give birth to a living baby, but the fear was too great. I was so scared because the delivery had been so traumatic and opted for a cesarean section. If only the baby was there, healthy and well, I did not care how.

And there he was, delivered at 38 weeks, our sweet baby boy with large blue eyes and good health—our present. I wish I could say that everything was fine after that. For the first year it was, and we enjoyed Faas. Of course, we also thought of Noud, and on his day, on May 1st, we would go do something fun with our family and we brought flowers to the cemetery where his ashes were scattered. We always honor his day.

Sam Riedijk

A very important moment to each couple is when the clinical geneticist tells them about genetic test results. To most couples, this moment of receiving the recurrence risk (estimation) feels like a decisive moment. Even though, regardless of the actual number, nearly all couples I met tried for another pregnancy.

There also seems to be a myth that couples should finish grieving before getting pregnant again, and that it is psychologically unfavorable to try to replace your lost baby with a new one. In my experience, if couples would have to wait until they were no longer grieving, they would be waiting for a long time. Also, I am yet to meet the couple that succeeded at replacing their baby. To most women the next pregnancy is the only comfort that actually helps them to bear the pain. The joy and hope brought by the next pregnancy provides some balance to the pain of loss. Partners may have more doubts about another pregnancy, fearing that if this would happen again, it might be too much to bear. Women, on the contrary, seem to fall prey to an overwhelming and uncontrollable desire to get pregnant again. This is another source of misalignment between women and partners. I consider women doubting their sanity to women designing ingenious traps to get their men home for their ovulation and everything in between as quite normal.

Once the family is complete, the focus of attention can shift from reproduction to child rearing. I have had the opportunity to work with some couples years after the pregnancy loss and have gained insights in the long-term dynamics of child loss after prenatal diagnosis. To nearly all couples, their lost child has grown into a full member of the family. The child often has a place in the house in some symbolic form, or a picture or ornaments. The child may have a name, and most couples I have worked with have told their children about their brother or sister, who was too sick to live.

Frederike Dekkers

The main recommendations are as follows:

- Provide guidance when parents are pregnant again.
- When parents are pregnant again, talk about the importance of bonding with the new child and refer to a psychologist if you have doubts about the quality of bonding.
- Acknowledge the difference in mourning between mothers and their partner.
- Acknowledge how hard it can be for couples to deal with their network, where many friends and family around them get pregnant and have a baby.

References

Campbell-Jackson L, Bezance J, Horsch A: "A renewed sense of purpose": mothers' and fathers' experience of having a child following a recent stillbirth., *BMC Pregnancy Childbirth* 14(19):423, 2014.

Fraiberg S, Adelson E, Shapiro V: Ghosts in the nursery. A psychoanalytic approach to the problems of impaired infant-mother relationships, *J Am Acad Child Psychiatry* 14:387–421, 1975.

Gravensteen IK, Jacobsen EM, Sandset PM, et al: Anxiety, depression and relationship satisfaction in the pregnancy following stillbirth and after the birth of a live-born baby: a prospective study, *BMC Pregnancy Childbirth* 18:41, 2018.

Kerns KA, Brumariu LE: Is Insecure Parent-Child Attachment a Risk Factor for the Development of Anxiety in Childhood or Adolescence? *Child Dev Perspect* 8:12–17, 2014.

Markin RD: "Ghosts" in the womb: A mentalizing approach to understanding and treating prenatal attachment disturbances during pregnancies after loss, *Psychotherapy (Chic)* 55:275–288, 2018.

O'Leary J: Grief and its impact on prenatal attachment in the subsequent pregnancy, *Arch Womens Ment Health* 7(1):7–18, 2004.

CHAPTER 11

Organizing support

Jane Fisher
Antenatal Results and Choices (ARC), London, United Kingdom

Organizing support for pregnant couples facing prenatal diagnosis of congenital anomalies. The experience of UK charity Antenatal Results and Choices (ARC)

The origins of Antenatal Results and Choices

Antenatal Results and Choices (ARC) (Fisher, 2008) started life as a registered UK charity under a different name. In 1988 "Support after Termination for Fetal Abnormality" (SATFA) was established "for the purpose of giving individual and group support to women and their families, who have undergone a termination of pregnancy for fetal abnormality. It is intended that SATFA will establish a national network of self-help support groups as well as work with healthcare professionals in the improvement of practice and services to women during their time in hospital and discharge into the community."

It is important to consider the landscape of antenatal testing when SATFA was founded. In the 1980s, there was no standardized antenatal screening; what was offered depended on clinical interest in individual hospitals. Women were encouraged to have amniocentesis if they were over 35 or due to their family history (Green, 1995). If a diagnosis of significant fetal anomaly was made, the assumption was that the woman would terminate the pregnancy, put it behind her, and concentrate on the next pregnancy. This was a time when stillbirth was only just starting to be perceived as bereavement. For the women who did end the pregnancy, there was little consideration of the psychosocial sequelae as she had made an active decision to end the pregnancy.

> *You are the very person responsible for the loss; abortion implies the baby was unwanted. Rather than commiserating with your loss, people tend to say how lucky you were the abnormality was detected. With a stillbirth somehow it is acknowledged more — that a birth, labor and delivery has taken place... There are more likely to be pain-killers available, photographs if wanted, registration of the stillbirth, provision of funeral if wanted... None of these exist for terminations; the baby has no identity as such, so it is difficult to mourn and grieve and come to terms with.*
>
> ***Diagnosis of anencephaly at 22 weeks, SATFA newsletter, December 1985.***

Prenatal Genetic Counseling
DOI: https://doi.org/10.1016/B978-0-12-823329-0.00014-3

A group of bereaved couples and enlightened health-care professionals recognized that neither the practical nor emotional needs of couples were being met and founded SATFA to help address this. The first services that were set up were a helpline and local support groups run by bereaved couples to enable couples to acknowledge and share the experience of their loss, give themselves permission to grieve, and start to rebuild hope.

The fact that health professionals and bereaved couples were involved in SATFA's conception set the tone for what the charity would become. There was discussion among the charity's management committee in its early days as to whether it should be a charity just for bereaved couples or should also aim to support health professionals. The final vote was to do both. This meant that early on, SATFA developed training programs for staff in how best to manage care around termination for fetal anomaly. This is in line with Frederike Dekker's recommendations about supporting health care professionals . Professionals were also crucial in signposting women and couples to SATFA. They provided expert input into the SATFA "handbook" that was first published in 1990 (ARC, 2017). This booklet was written to sensitively prepare women and partners for what might be ahead if they opted for termination and to inform them of some of the choices they might have within the process.

SATFA received funding from the Department of Health to print the handbook on termination and to distribute it to all maternity units in hospitals in England, Scotland, and Wales. This helped raise the profile of the charity and increased the number of calls on its national helpline. Soon expectant couples were contacting the SATFA helpline after a prenatal diagnosis, looking for help in making the decision about the future of the pregnancy. To reflect this shift in contact, the charity changed its name to Support *around* Termination for Fetal Abnormality in 1992. As the 1990s progressed and more antenatal screening was introduced, the helpline team began dealing with calls about testing more generally, from women with questions about biochemical screening to those wondering whether to have amniocentesis or dealing with unexpected ultrasound findings.

Support after termination for fetal abnormality becomes Antenatal Results and Choices

In 1998 the Trustees of SATFA decided to officially widen the charity's remit to mark the change in contact and changed the name to ARC. Some women and couples who had sought support from SATFA were disappointed that the charity name no longer contained "termination." They worried that this might be seen as reinforcing the taboo and suggest that "termination" should be something hidden and unspoken. The fact was that ARC wanted to ensure that all those who needed nondirective information and support around all aspects of prenatal screening and its consequences had a place to come. However, ARC has continued to provide specialized help to those facing or having undergone termination as it is the only organization offering this service in the United Kingdom.

ARC maintained its partnership with health-care professionals and expanded its work with researchers and policy makers. ARC updated its handbook for women/couples ending a pregnancy for fetal anomaly and launched a handbook for those continuing in 2003.

How Antenatal Results and Choices works

The Antenatal Results and Choices national helpline

The core service of the charity has always been its national helpline run from its London office. The members of the small helpline team are not medically trained but have an intensive induction before dealing with contacts. The aim is to help staff build a knowledge base of the context and process of prenatal screening and diagnosis in the United Kingdom, and an understanding of when and where to signpost contacts for expert clinical advice and information about living with the condition that has been diagnosed. Perhaps even more importantly, helpline staff have skills and training in effective empathetic listening and sensitive communication. All have the capacity to be with people in acute distress and to contain those experiencing high levels of anxiety. The ARC team deals with approximately 6000 contacts a year by phone, email, and live chat. ARC also sees couples and women who have access to London for face-to-face sessions.

Most people who contact ARC have been given details by a health-care professional; the second most common way of finding ARC is via the Internet, either through recommendation from a social media or networking site or via a search engine. When we audit the service (this was last done in 2018), the consistent message is that people contact ARC for three main reasons. The first reason for contact is that couples want an independent source of nondirective information. This is not because they doubt what they are told by their clinical team, but they need help to understand some of the more technical information and be made aware of other sources of appropriate information, such as the nearest center of excellence for fetal medicine or other support organizations that may have expertise in life with the condition that has been predicted or diagnosed. Couples do not always want to bother busy clinicians with questions, and they value speaking to someone who has knowledge of what they are dealing with but at the same time is independent, impartial, and not connected to them either personally or professionally.

> *I contacted ARC during our darkest moment when we were on the point of having to make a very difficult decision, a decision that thankfully we were never called on to make. The information, help and support that your organization provided has greatly aided both my wife and I to get through everything that was thrown at us during the pregnancy.*
>
> ***E-mail to ARC, August 2016.***

The second benefit users of the helpline mention is that they are offered the time they might need to talk through possible outcomes of decisions they may be faced with after a prenatal screening or diagnostic result. This is in congruence with "scenario-making" that was mentioned earlier by Wildschut et al., Dekkers, and Riedijk. They can take as much time (and as many calls or emails) as they want to help them work out a way forward that is most appropriate in their personal circumstances. They can discuss what is happening and start to consider what it might mean to them. Issues couples are facing can be ethically sensitive, so it can be of comfort for them to know they have the opportunity to work things through in a safe confidential space.

Third, service users tell us they appreciate the acknowledgment and understanding of the psychological load of the circumstances they are experiencing as a result of a prenatal test result. Many are taken aback by the intensity of their feelings when confronted with a diagnosis and struggle to find a way to decide how to proceed. While ARC does not purport to offer either medical advice or therapeutic counseling, the helpline team's calm measured approach can facilitate the woman and couple's capacity to enact agency.

> *We were given the diagnosis over the phone, we were told there was no one available for us to talk to about it and could we ring back when we'd decided what we wanted to do. It was at this point, desperate for help; my husband came across the charity ARC (Antenatal Results and Choices) and rang them instantly. This was the best thing we could have done; just feeling that what was happening to us mattered to someone was so important. We rang them a number of times through the weeks that followed. They don't tell you what you should do, they support you with kindness and balanced information so you can make the decision that is right for you.*
>
> ***Social media post to encouraging fundraising event run by couple supported by ARC 2019.***

To sustain a high-quality service the ARC helpline team follows a "turn and talk policy." After a difficult call, immediate peer support is always available, which is overseen by the Director of the organization. All those who take helpline calls have monthly external supervision to help manage the psychological demands of the work.

How the changing sociopolitical climate regarding prenatal diagnosis affects couples

The last decade has seen rapid developments in prenatal screening and diagnostic technologies. We have seen more sensitive genetic diagnostic techniques such as microarray enter clinical practice and are on the cusp of the more widespread introduction of prenatal whole genome and exome sequencing, as can be read in the chapters by Diderich et al. and Lewis et al. At the same time, there have been improvements in the application of prenatal imaging via ultrasound and magnetic resonance imaging. Cell-free DNA testing has increased detection rates for aneuploidy and is providing a safe diagnostic tool for certain single-gene disorders. These developments have provoked interest in the literature about the ethical implications of presenting expectant couples with an increasing amount of information about their developing fetus (Dondorp et al., 2015). The concerns include the challenge of communicating results that may be difficult to interpret or have uncertainty regarding outlook. Some commentators fear we are creating a "soft eugenic" climate, with expectant couples pushed toward seeking the "perfect" baby (Dixon, 2008).

Some people living with disabilities and genetic conditions and their advocates have started to question whether the pursuit of prenatal diagnosis and the accompanying offer of termination is inherently "ableist (Boardman, 2014)." Couples who terminate are seen by some as valuing a potentially disabled child less highly than one without disability and thus perpetuating discrimination. These arguments are not new, but the last decade has seen them extend beyond academia and take hold in the UK mainstream media and on social media platforms.

The most vociferous debates in the United Kingdom as well as Denmark and the Netherlands have centered on prenatal testing for Down's syndrome and specifically whether the introduction of cell-free DNA screening will lead to a significant reduction of trisomy 21 live births. Some campaigners lobbying against cell-free DNA screening being implemented into the National Health Service in the United Kingdom have targeted ARC with accusations that it colludes with the medical community in perpetuating myths about what life with Down's syndrome can be like and assumes couples will terminate when a diagnosis is confirmed.

In the last 20 years the numbers of terminations for Down's syndrome in England and Wales have remained quite consistent at approximately 90% (Morris et al. 2016; UK government, 2017). We know from our support work with women and their partners and health-care professionals that expectant couples do not take the decision to end their desired pregnancy lightly. The choice they make is not about devaluing people living with the condition but about what the prospect of having a child with Down's syndrome means to them in their own *individual* lives. It is of course crucial that those considering testing for Down's syndrome or facing a diagnosis of the condition have access to accurate and up-to-date information on what life with Down's syndrome can be like.

The current "politicization" of Down's syndrome adds a further layer of difficulty to an already fraught process for expectant couples. While they may often be open to accepting a child with a disability into their lives, the socioeconomic reality they face in contemporary Britain can appear less than welcoming. Disability inclusivity has improved in this century through societal and legislative changes, but there remain major challenges in ensuring that people living with disabilities and their families have access to the resources they require (https://www.equalityhumanrights.com/sites/default/files/being-disabled-in-britain.pdf.). It is important to take the lead from expectant couples and do everything possible to help them to take decisions that, taking account of the reality of their individual circumstances, they are best able to live with.

Conclusion

Prenatal testing technologies continue to develop. We have moved swiftly in the last decade from karyotyping to whole genome and exome sequencing, exponentially increasing the amount of information available to expectant couples about their developing baby (though not accompanied by an increase in certainty of postnatal outcomes). What has not changed over the same time is the psychological impact of the news of a confirmed fetal anomaly. For couples and families the medical "facts" of any diagnosis will be refracted by their own life experience, psychology, and their relationship to the pregnancy and to each other.

Carefully coordinated individualized and compassionate care from health-care professionals through prenatal diagnosis is paramount. However, the experience of ARC has shown that there is also value in the availability of an independent but complementary service, offering nondirective information and support to help women and partners negotiate prenatal diagnosis of fetal anomaly, which is likely to be one of the most challenging times in their family lives. While there may be sociocultural differences in the prenatal testing arena outside the United Kingdom, as seen from this handbook (see the chapters 1 and 7), the core practical and psychological support needs of couples will be similar, so the establishment of comparable services in other settings might be considered.

References

Accessed 09/02/2020, at https://www.equalityhumanrights.com/sites/default/files/being-disabled-in-britain.pdf

ARC: *A handbook to be given to parents when a fetal anomaly is diagnosed*, 6th ed., London, 2017, ARC.

Boardman FK: The expressivist objection to prenatal testing: the experiences of families living with genetic disease, *Soc Sci Med* 107:18–25, 2014.

Dixon DP: Informed consent or institutionalized eugenics? How the medical profession encourages abortion of fetuses with Down syndrome, *Issues Law Med* 24:3–59, 2008.

Dondorp W, de Wert G, Bombard Y, et al: Non-invasive prenatal testing for aneuploidy and beyond: challenges of responsible innovation in prenatal screening, *European Journal Of Human Genetics* 23:1438, 2015.

Fisher J: Termination of pregnancy for fetal abnormality: the perspective of a parent support organisation, *Reprod Health Matters* 16:57–65, 2008.

Green JM: Obstetricians' views on prenatal diagnosis and termination of pregnancy: 1980 compared with 1993, *Br J Obstet Gynaecol* 102:228–232, 1995.

Morris JK, Grinsted M, Springett AL: Accuracy of reporting abortions with Down syndrome in England and Wales: a data linkage study, *J Public Health (Oxf)* 38:170–174, 2016.

UK government, 2017. 2020, at: https://assets.publishing.service.gov.uk/government/uploads/system/uploads/attachment_data/file/821449/NCARDRS_congenital_anomaly_stats_2017.pdf

CHAPTER 12

Staying young; student reflections

Diana Zavyalova[1] and Jasmijn E. Klapwijk[2]
[1]Minor "Genetics in Society", Erasmus Medical Center, Rotterdam, The Netherlands
[2]Department of Clinical Genetics, Erasmus Medical Center, Rotterdam, The Netherlands

While compiling this book and reading through the patients' experiences, we came to realize the importance of "staying young" as professionals. That means remembering our first cases; our first conversations with our first patients; their heartbreaks and triumphs—all of which we were rooting for and living through together with them. Therefore we suggest that as a professional in the field, surrounded by experience, usual responses of choice, and average results, you ask yourself the following questions: Do I remember the emotions I lived through when I first started this journey? Am I able to combine my expertise with my own human perspective?

In this chapter, we outline the common perceptions of those that are not yet bound by their own experience: young students from various backgrounds. They have discussed fetal genetic abnormalities and such patients experience directly with Lara and Iris. They heard the stories about Mae and Noud in person, and, touched by the raw emotion behind the science of it all, are represented through a collective experience summary, outlined in this chapter for all of us to remember what it was like to be in their shoes.

Phase 1: the discovery

During this first stage of Lara and Iris's stories, both described how they experienced their initial confrontation with the possibility of an abnormality in their unborn child. They described the procedure of the ultrasounds, the emotions they felt when bad news was shared, as well as the emotions and thoughts on how that was shared with them. Many students reflected on two insights: the change in their own understanding of pregnancy and its duration, and the care pregnant women receive during the ultrasound procedure that leads up to follow-up testing.

The primary insight bloomed from the shared perspective of Lara and Iris on the 20-week ultrasound. Many students confirmed that naturally, the key component of the appointment they had in mind and envisioned for themselves was the disclosure of the sex of the baby. For many, it came as a surprise that not only do the sonographers look for features other than the sex, but this is also exactly the moment when parents

Prenatal Genetic Counseling
DOI: https://doi.org/10.1016/B978-0-12-823329-0.00012-X

hear about structural fetal abnormalities if there are any. Many students were forced to face the reality of the procedure and realized that the practice of ultrasound carried much more severe and significant connotations than they expected.

> *Without this [conversation with Lara and Iris], it is entirely possible that I would have been the ignorant person to be anxious for the 20-week ultrasound, because of the sex of the baby. My headstrong, inflexible approach to parenthood didn't allow me to see much of the bigger picture. This experience has slammed that bigger picture in my face. In the grand scheme of things, this has made me realise that when you are utterly focused and convinced of something, it can narrow your perspective on the matter, which is something I should keep in mind professionally as well.—Sander Matthijssen.*

> *I was particularly struck by how both Iris and Lara saw it as a blessing that some couples only focus on the excitement of the sex determination at the twenty-week ultrasound. This impacted the way I viewed my own possible future pregnancies, as well as the way I interact with those who are pregnant now. [...] Especially in a society where pregnancy announcements and gender reveals can be huge and happy events, Iris and Lara made me realize that it is very important to remain sensitive to those who experience difficulties in pregnancy. In my future profession as a public health specialist or researcher, it is not very likely that I will deal with pregnant women directly. However, Lara and Iris taught me how, for every individual you encounter, there might be more to them than meets the eye.—Miriam Sijtsma.*

> *Everyone said something different, sometimes it wasn't clear what to expect. Iris told [us] she would've wanted not to be reassured when there was a risk of the pregnancy going wrong. What I will take from this is paying attention to what the patient wants to hear and be more careful whether to lead someone on with unjustified optimism.—Liz Tupker.*

Phase 2: decision-making

In the next stage, Lara and Iris had invasive testing to determine whether there was a genetic cause for the detected structural abnormalities. Along with this came the decision-making process of whether or not to end the pregnancy.

This was a moment that made the students question their own understanding of the abnormalities' severity: Would I have done the same if I were in their situation? What are the boundaries that allow one to continue a pregnancy or to terminate it?

> *Listening to Lara's and Iris' story, however, made me realize that there are very understandable and reasonable reasons why one would end their pregnancy, besides just not wanting to have the child. Even though I still believe that every embryo already possesses a "soul" and is a sentient being that deserves a life, I no longer see the whole topic as solely black and white. [...] This insight is especially beneficial for me personally, as it broadened my horizon tremendously and let me reconsider my somewhat gridlocked opinion.—Annabelle Bodmann.*

> *Before I had the conversation with Lara and Iris, I always thought that terminating a pregnancy would be a very practical decision. The new insight that I got is that [...] it's also a very emotional decision to make. [...] I always thought to myself: 'If I find out that my child is*

severely handicapped or sick, I'm going to terminate the pregnancy because I don't want to live the rest of my life taking care of a handicapped child'. Now I know that making such a decision is way harder than I thought. The emotional stress and trauma that Iris and Lara got from terminating the pregnancy is enormous, even after all those years.—Mieke Wijsman.

Some also emphasized the importance of proper care throughout the decision-making process. As Lara and Iris mentioned in their talk with the students; having a health-care professional (clinical geneticist and counselor) nearby and ready to communicate as soon as something becomes known, thus eliminating vagueness and further shock, provides couples in a similar situation with an anchor that grounds them and reassures them of the next course of events.

The importance of patients [. . .] being cared for by the clinical geneticist but also by other people being involved in this process, such as the sonographer, became apparent throughout both of their stories. Both Iris and Lara explained that the uncertainty of not knowing how to proceed with their pregnancy due to waiting for further information [. . .] posed the biggest challenge. [. . .] However, both women mentioned that they felt comforted by the emotional support provided by their clinical geneticist, who gave them the impression of being their contact person whenever they needed her.—Tim Schäfer.

Phase 3: the delivery

This brings us to the next and, perhaps, the most shocking phase of the experience: the delivery. In their stories, Lara and Iris described the procedure quite similarly up until the point of delivery. While Lara embraced having given birth to a child with abnormalities by taking her daughter home, dressing her, taking photographs, and giving her a name, Iris wanted to leave that day in the past and tried to escape from this traumatic experience by distancing herself from her child as much as possible.

It does not come as a surprise that this was the part that most students struggled with. Many were astonished by the described environment at the hospital, and almost all students were taken aback by completely different approaches Lara and Iris had toward the delivery and their babies.

The part [. . .] that stuck with me the most is the delivery of the deceased babies. In the past, I was aware women could decide to terminate a pregnancy when there are fetal anomalies present, but I never thought about the actual process. I just assumed that the woman would be under anesthesia while a professional would remove the fetus. It was very shocking to hear they had to deliver the babies vaginally [. . .]. I thought it was highly inconsiderate [that] they had to deliver the babies in the same corridor as all of the women with healthy babies. [. . .] I get that it was easier to let them deliver the babies in the same corridor as all of the other pregnant women, since all the right equipment and staff was nearby. On the other hand, I also saw this decision had a big impact on the feelings of the women.—Mirthe van der Meijden.

[. . .] [Iris] didn't want to touch or hold the foetus, didn't want to name it, and she was convinced that this was only a failed attempt at having a baby [. . .]. I remember feeling a sense

of agreement; I felt that this is also the way that I would like to handle the situation [. . .]. This is especially the case since I remember being quite shocked at how Lara behaved in her situation. [. . .] I caught myself thinking 'this is not the right way', and when Iris told us [about] her approach, I was somewhat relieved. Sadly, as Iris' tears appeared, my relief was washed away.—Sander Matthijssen.

There can be different reactions after the stillbirth of a child. Somehow I had expected that mothers would have liked to see their child, hold it and say goodbye to the baby [. . .]. Apparently, there can be a big difference between mothers and I think it is very personal what someone likes [in this situation].—Sylvia Tromper.

Many students were also able to draw a parallel between the core of their emotions and what it meant for them as future professionals with the premise here always being personalization for each couple.

Being aware of couples varying in their reactions to [TOP], it seems reasonable to offer individualized support—for example, in counselling sessions—rather than helping every couple in the same way. Additionally, clinicians should be prepared for [confrontation] with different coping strategies and avoid [imposing] certain ideas or perspectives upon patients, as there is no correct reaction to [TOP].—Tim Schäfer.

I'm wondering what the best way is to help mothers process the loss of their [babies]. If someone pretends or says it doesn't have to be a big deal, how will this affect the mother? In the long run, what is the best way to help the mother to process losing the baby right after the birth? To me, it seems to be a really difficult situation and I really cannot imagine how intense it would be to have just given birth to a baby without a beating heart.—Sylvia Tromper.

Phase 4: after ending the pregnancy

The importance of the presence of such individualized support around those couples who are mourning through this entire experience brings us to the next part in the termination process: grief and its management, both right after birth and many years later.

One of the main insights that students had was the fact that grief comes in different ways for everyone. Where one person prefers to work through it thoroughly, involving as many people as possible for emotional support, another would rather grieve quietly and privately, preparing themselves for years to open up to the traumatic experience.

Lara and her partner [. . .] had a name for [their daughter], a basket to bring her home in, Lara made a nice blanket to wrap around her [. . .] and when they came home, they had a small celebration to kind of acknowledge that their daughter was born. On the other hand [are] Iris and her partner who didn't give their son a name when he was born and didn't take [him] home. There are a lot of differences between Iris and Lara, [which] made me realize that people respond to certain events differently [. . .]. And that different ways are not good or bad but their own decision. [. . .] But because people make different decisions, there should also be different ways to help [them while they are] going through this process and to support them [in the aftermath]. [. . .] If this happens, maybe people [would] feel more comfortable explaining their decision to [their] family, so [that] they get new insights too and make it less of a taboo to talk about [. . .].—Iris Clavan.

Iris wanted to forget about the whole situation. She didn't give her baby a name. She preferred not to talk about it. Others made her feel like she shouldn't make [it] a big thing. [. . .] Yet years later she was actually bottling up her feelings of trauma. [. . .] Someone [had recommended it] and then she decided to visit a genetic counsellor. [. . .] It changed everything. The acknowledgement of [the] loss created an opportunity to give it a place. I wondered if [Lara and Iris] could ever be free of the feeling of guilt. They told me they can't.—Liz Tupker.

Most surely, these insights were translated by the students onto the bigger picture of the professional field; how important it is for genetic counselors to stay aware of the differences in each family and be able to provide relevant support that the family can benefit from in the long term.

[. . .] without properly dealing with this "big" thing out in the open, it grows within you, hidden, until you can no longer contain it. Thankfully, due to Dr. Riedijk's counseling, Iris was able to find peace within herself and realize the things that she would have done differently. This shows how important it is for counsellors to bring up the events that their patients buried long ago but never dealt with; how important it is to not shy away from making someone feel uncomfortable for the sake of helping them achieve peace in life and if not fully forgive themselves, then at least think about their reaction to the pregnancy loss in a different, self-sparing manner.—Diana Zavyalova.

Years after the event, [Iris] still found herself reacting emotionally to triggers. She only came to terms with it once she started to reflect on what she had experienced. She now accepts her lost baby as a part of the family and shares her story. This taught me that different coping strategies can be more or less effective. However, in [counseling], I believe it is important to guide a patient in little steps instead of forcing him or her to follow a certain strategy. Everyone has their own pace, and Iris might not have benefitted from another coping strategy shortly after going through the termination of pregnancy.—Maylin Grabbenstein.

Reflecting on support needed throughout the process

The necessary distinction between each couple's experience with the termination of pregnancy brings us to the final, all-rounding point within students' reflections. The open and raw conversation with Lara and Iris allowed them to see for themselves what kind of support needs to be organized to help prospective parents deal with the very difficult situation that they are facing. It allowed them to see that they should not be alone. They should not have to hide this part of themselves deep within them. Instead, they must be helped as much as possible before the grief acquires a perpetual shape and becomes a major part of their lives.

I think that as a professional, I can take these stories with me to see how I can best support people in the decision of what to do after giving birth. [. . .] I think the most important thing is that the mother does something that she feels good about and regrets as little as possible later on. And if a mother regrets something, I want to approach [her] with an open, non-judgmental attitude.—Sylvia Tromper.

Both women mentioned that they felt comforted by the emotional support provided by their clinical geneticist, who gave them the impression of being their contact person whenever they needed her. Additionally, Lara said that one of the nurses helped her get through the process of delivery by strongly conveying the feeling of taking his time in order to be there for her. Consequently, clinicians should recognize that a well-established cooperation with patients results from initially focusing on the patient's needs. Moreover, the communication between both parties should not only include the medically relevant questions from the clinician's side but also leave space [and] time to make the patient feel at ease.—Tim Schäfer.

Discourse as a valuable learning experience

All in all, most students agreed that the conversation with Lara and Iris allowed them to broaden their horizons not only from a personal perspective but also as future professionals in the field. While not all of them plan to become clinical geneticists or genetic counselors, all still reflected on the beneficial impact of such discourse being present in the medical environment as they remind professionals and medical researchers that the human aspect of prenatal genetics is critically important; that while there are only two outcomes in the end—termination of pregnancy and its continuation, each case is drastically different from another and, therefore, asks for tailored care and advice.

While the discourse often prevails in such conversations, resulting from the initial shock of the circumstances, we must also remember that human resilience is stronger than grief. Without the ability of human beings to recover and move on from loss, tragedy, and other traumatizing and heartbreaking events, society as a whole would fall apart. More specifically, mothers who lose their fetuses to genetic diseases would not be able to recover well enough to have more children, therefore not participating in the creation and upbringing of society in its everlasting, cross-generational form. Therefore remembering and honoring human resilience—as individuals and professionals of this field—has been, is, and will always be our lifeline.

When I measured Iris and Lara's resilience to mine during the session, it seemed unimaginable. Talking about it to a classroom full of quiet strangers, being able to discuss it without feeling like the grief belongs to you and only you – it blew my mind and left me looking for words as I tried to explain the feeling to them in person. I can see how realizing the existence of such a level of resilience can be beneficial for a genetic psychologist's understanding of his/her patients. Of course, everyone deals with pregnancy loss differently – Lara and Iris are prime examples to this – and not everyone can deal with such a matter healthily, or on their own, or at all. However, seeing such examples, learning from them and appropriately using that experience for the following patients, appears to be sensible and useful in practice.—Diana Zavyalova.

I think it is important to be confronted with tough subjects like this, because I feel like it is often forgotten that the decisions we make from a practical viewpoint can have a tremendous psychological impact on the patient. When you constantly learn about theories and the most practical way to solve a tough situation, you distance yourself from the situation and forget about the psychological consequences. The earlier students are aware of these effects, the bigger the chance they will take this into account while making medical decisions.—Mirthe van der Meijden.

Something else I learned from the conversation with Lara and Iris is how important it is to give sufficient weight to the act of ending your pregnancy. To put it in Iris' words: "Make it big, because it is big." Ending a pregnancy, especially at such a late stage, is not something you just do before you continue normally with your life. Such a decision comes with a lot, emotionally as well as mentally. Both women taught me how important it is to give yourself the room to cry, grieve, cope, and talk about it.—Annabelle Bodmann.

Conclusion

Many students reflected that they had never thought about the possibilities of an unsuccessful pregnancy as much as they started to after the conversation with Iris and Lara. That is why, for us as professionals, it is important to remember that this is not an accepted standard or a common occurrence that people are used to. No class, amount of papers, or book chapters could have prepared us for the amount of heartbreak in Lara and Iris's stories. The sheer volume of grief and heart-wrenching humanity blew the students away and ensured that everyone, every single person in the classroom—even those without any experience with losses to a genetic disease or a pair of ovaries—could feel those stories so deep inside them that some were brought to tears, while others spent the rest of the day aloof and trying to comprehend such tragedy. This experience considerably impacted these "young" professionals, and we conclude that years later, saturated with experience and professional rigor, these shared stories and experiences should stay in our peripheral vision and remain there until the end of time.

Appendix: Concluding remarks

Writing this book, we were driven by the wish to make our lessons learned available more broadly. We did not think of the taboos surrounding this topic, we only thought of all those couples who need our care in the most challenging times of their lives. Writing this book gave us the opportunity to piece together where we are at in the domain of prenatal genetics, what couples experience, and what we have learned to offer them in terms of (after) care and decision-making support. It made us realize how fast this field is developing, adding challenges, such as uncertainty, and also how timeless the needs of couples and parents are when faced with prenatal anomalies. Couples are often unprepared and overwhelmed when informed of fetal anomalies. They need time and information, as well as support of healthcare professionals to grasp, understand, and decide about what course to take. Although people differ, many of the themes we encounter in prenatal genetics can be found across the globe, such as the pain of having to decide about the pregnancy, the desire to give one's child a life worth living, and the grief when fetal anomalies. There may be differences in contexts and cultures, but the essence of being human and what it means to become a parent show more similarities than differences. One thing is certain to us, providing good care to couples in the context of prenatal genetics is care that comes from the heart. In the safe space of the relationship, we can offer our patients the opportunity to gain psychological grip. We follow their lead and provide them with a road map. As we were studying the cases that came to mind for this book, we realized that we are far from perfect. This is in contrast to the responsibility we feel toward our patients. It is important to share how we care for couples, what we have learned to work well, and also what did not, and where we were in doubt. When caring from the heart, we make mistakes, but we learn from them and next time we will do better. Caring for each other and ourselves implies showing as much compassion to ourselves as we show our patients. This book is our contribution to caring for our patients and each other.

Acknowledgment

The contribution of Frederike Dekkers would not have been the same without the time and effort invested by Attie T.J.I. Go, Luuk Stapersma, Alex J. Eggink, & Elisabeth M.W.J. Utens. Many thanks for their inspiring work and input in the research of parents' preferences for psychosocial care before, during and after a termination of pregnancy.

Index

Note: Page numbers followed by "*b*" and "*f*" refer to boxes and figures, respectively.

Q

R

S

T

9780128233290